The Last Extinction

To Alexis
love
Jan

The Last Extinction

Second Edition

edited by Les Kaufman and Kenneth Mallory

Published in association with the New England Aquarium

The MIT Press
Cambridge, Massachusetts
London, England

This book was printed and bound in the United States of America.

The Last extinction / edited by Les Kaufman and Kenneth Mallory.—2nd ed.
  p. cm.
"Grew out of a public lecture series entitled 'Extinction: saving the sinking ark,' held in Boston, Massachusetts, at the New England Aquarium during the fall of 1984"—Pref.
"Published in association with the New England Aquarium."
Includes bibliographical references and index.
ISBN (invalid) 0-262-11179-1 (HC).—ISBN 0-262-61089-2 (pbk.)
  1. Biological diversity conservation. 2. Extinction (Biology) 3. Man—Influence on nature. 4. Nature conservation. I. Kaufman, Les. II. Mallory, Kenneth. III. New England Aquarium Corporation.
QH75.L36 1993
333.95—dc20   93-12575
                                                                              CIP

Contents

Preface

As you will read from the appendixes to the chapters of the original edition of this book, this second edition of *The Last Extinction* reflects the many changes that have occurred over a period of seven years. On the downside, the three remaining and captive dusky seaside sparrows chronicled in the Williams and Nowak 1986 chapter "Vanishing Species in Our Own Backyard" are sadly no more. On December 12, 1990, the U.S. Fish and Wildlife Service declared the dusky seaside sparrow officially extinct, a victim of politics and neglect. If we include the continued destruction of the world's rain forests, the demise of the dusky is just one of tens of thousands of extinction stories that have occurred in the past seven years.

On the plus side, there is a new and stronger global awareness of the issues the first edition of this book took pains to address: the ongoing reality of mass extinction, destruction of biological diversity caused by the Earth's ultimate predator, *Homo sapiens.* As this book goes to press, newly elected Vice President Al Gore is sharing his call for environmental advocacy in a book that is the fastest selling hardcover environmental book of all time, with its first year sales of 250,000 copies outpacing previous best first year sales (Rachel Carson's *The Silent Spring*) by 150,000 copies. In *The Diversity of Life,* Harvard University's Edward O. Wilson has presented the options he sees for our planet in his most impassioned and clearly reasoned argument yet.

The positive side of environmental awareness isn't confined to paper and pen (computer screen and kilobyte). Here in New England one of the North Atlantic coast's richest mines of biological diversity, Stellwagen Bank, is now a national marine sanctuary. The Belize barrier coral reef is the recent recipient of support from the World Bank, the United Nations Development Program, and the United Nations Environment Program to create a similar sanctuary off the coast of Central America. The black-footed ferret of America's Great Plains, poised on extinction, has now recovered through captive breeding to a point where they have been reintroduced to the wild (see the appendix to chapter 5).

The New England Aquarium, organizer of the first and second editions of this book, has changed, too. We are in the middle of plans to expand and diversify our aquarium with spectacular new building additions for our present site on Boston Harbor's Central Wharf. The driving force behind our plans is a newly crafted mission "to increase understanding of aquatic life and environments, to enable people to act to conserve the world of water, and to provide leadership for the preservation and sustainable use of aquatic resources."

The world's aquariums offer a real hope for being a force in preserving aquatic diversity, but it's a new idea and a difficult challenge for museums that have traditionally focused on family entertainment and passive education. One of the greatest contributions we can make to conservation may be to make our collections centers for research, just as preservation was the goal of museums in the past. Institutions such as the Columbus, New York, Indianapolis, and Chicago aquariums have already played a major role with captive breeding programs and conservation efforts in the field. Many new institutions such as the Tennessee State Aquarium in Chattanooga and the Osaka Aquarium in Japan were designed with strong conservation roles in mind. The ultimate contributions of so many of these institutions are yet to be realized.

In addition to the new cover photo of the North Atlantic right whale, the most endangered whale in the ocean today, this second edition contains a new chapter called "Sharing the Earth with Whales," by environmental writer and consultant Norman Myers. Dr. Myers's timely chapter addresses the critical issue facing biodiversity as we approach the twenty-first century—habitat destruction, which for whales involves over 70 percent of the Earth. The changes in the Amazon are reflected by new information added as paragraphs and updates throughout Ghillean Prance's chapter "The Amazon: Paradise Lost?" while the other four authors add new perspectives with chapter ending appendixes. Only David Ehrenfeld's essay remains as it was in the first edition, standing as it does as a timeless and eloquent plea for environmental stewardship. The final section of the book offers an updated list of the many organizations devoted to protection of the Earth.

Preface to the First Edition

This book grew out of a public lecture series entitled "Extinction: Saving the Sinking Ark," held in Boston, Massachusetts, at the New England Aquarium during the fall of 1984. Our hope for the series and for this book is to awaken the general public to the issues underlying the notion that we must prevent a "last extinction." Beyond ringing an alarm, we hope that you will realize that a mass extinction is in progress but that it can be postponed indefinitely; that doing so will require a rearrangement of your priorities; and that your decisions will affect the face of this Earth, including the economic and spiritual welfare of your children. This book is to help you make your own decisions now.

Many people know that we are rapidly losing the diversity of life on Earth, but they are not doing anything about it, either because the idea is only an abstraction or because they think that the problem is too vast for individuals to have any effect. For these people we offer a bit of bumper-sticker wisdom, revealed to us in a traffic jam: "Think Globally, Act Locally." This book is an attempt to bring extinction down to earth, to help you understand what it is, and what it isn't, so you can help to reverse it.

There are several new books about extinction with more on the way, but this one differs from those of our colleagues. As employees of a major public aquarium that hosts about one million visitors a year, we have heard some pretty strange interpretations of what evolution and extinction are about. We are hoping to clarify a few of these misconceptions. We have noticed a polarity in the way that our visitors perceive scientists: some think of them as all-knowing gods, whereas others think they are a bunch of wafflers. For these people we have tried to make clear distinctions between the certainties and uncertainties of mass extinction, illuminating the uncertainties as much as possible. Other books emphasize the importance of tropical deforestation in the loss of species. This is justifiable, but most temperate zone inhabitants do not understand what jungles have to do with their lives. We have brought the issues home to everybody by reviewing case studies from both the temperate zone and the tropics and from both aquatic and terrestrial habitats. Finally, we have tried to present our case without becoming too strident. We may not have tried hard enough. The authors and editors of this book are too intimately familiar with the details of mass extinction not to feel great anger and sadness over them, and sometimes these feelings show through. We hope that our words will find as receptive an audience in print as they did in the lecture hall and that the book will be something more than a catharsis.

The first two chapters deal with the nature of mass extinctions and the role of people in causing this one. Les Kaufman, a marine ecologist, examines philosophical, scientific, and practical stumbling blocks to the conservation of biological diversity. David Jablonski, a paleontologist, explores the nature of extinction in the fossil record. Jablonski supports a view held by many scientists that we are presently experiencing a mass extinction, a major and catastrophic event in earth history with serious consequences for the quality, and perhaps even the continuance, of human life.

The next two chapters are case studies. Ghillean Prance, a tropical botanist, describes the situation in the Amazon rain forest, a rich tropical community sprawled across several developing nations. In comparison, Jim Williams and Ronald Nowak of the United States Office of Endangered Species assess the state of affairs in North America, whose familiar temperate communities are host to an overdeveloped world superpower. Both chapters emphasize the need to conserve habitats to safeguard the species dependent on them.

In the final chapters, two realists grapple with the future. Thomas Foose, Conservation Coordinator for the American Association of Zoological Parks and Aquariums (AAZPA), is among the principal architects of a global zoo ark known as the Species Survival Plan. Foose explores the limits of this last-ditch effort at captive preservation of species left homeless by habitat destruction in the hopes of reestablishing them in the wild once their habitats have been restored. In the closing paper, David Ehrenfeld, a biologist and conservationist, takes a long view of the kind of world we might be living in if we do, or do not, make our best efforts at preserving life's diversity.

At a glance, this book appears to have been assembled by two people. Somebody had to write it, though, and we are indebted to our five contributing authors for providing us with both a lecture and a book. The contributors themselves were brought to us by John Lowell and the Lowell

Foundation, which has provided us with generous support for two major public lectures series per year, as well as special programs for the ill and the handicapped. Mr. Lowell, in turn, was brought to us by the New England Aquarium, which sponsored the lecture series and supported our efforts to publish it. This book is a cornerstone in the aquarium's new program in aquatic conservation. We owe special thanks to David Ehrenfeld, who helped greatly in editing several of the manuscripts; to the Cousteau Society, especially Susan Richards and Jean-Michel Cousteau, and to Loren and Sue McIntyre, for their help with the photographs that illustrate this book. Finally, we are grateful to our wives, Jackie Liederman and Margaret Thompson, who lent much to the initial conception of this book and who by now must feel as if they have lived through several successive extinctions of their mates.

The Last Extinction

Why the Ark Is Sinking

Les Kaufman

Human beings are the most adaptable creatures that have ever lived on Earth. Reason and insight, the chief human talents, have given us the power to forge a world increasingly fit for our own comfort. We feed voraciously on all other manner of life, from whale to lily. We have no significant predators save a diminishing roster of infectious diseases. Of course, there is also the odd shark, crocodile, or lion, but they are disappearing even faster than the diseases. We carve the planet's surface into fields and streets, shopping malls and parking lots, with little regard to what was there before, because what we replace it with offers a more immediate, short-term benefit. Never before in Earth's history has such an abundant, aggressive, industrious omnivore at the peak of the energy pyramid comprised such a large portion of the living biomass. The prognosis is clear for the five to ten million other kinds of living things that share the Earth. They are in big trouble.

The world is host to two hundred nations and what amounts to five billion rulers. The strain of catering to so many separate interests is manifest in the flow of political, economic, and social crises. All living things are affected by these crises, but humanity as a species has thus far survived them and prospered. The bounds of human habitability include nearly the whole globe, whereas the entire liveable universe of other species can vanish overnight as one river is dammed or one hillside is laid bare. As we eliminate each species that stands in our way today, we lose any hope of having it back tomorrow. Life on the planet advances irreversibly, like a ratchet, toward greater impoverishment.

The ratchet clicks faster each day. Between the years 1600 and 1900, species of mammals and birds vanished at the rate of approximately one every four years. During the twentieth century, mammals and birds have disappeared at the average rate of about one species per year. The fossil record is too poor to provide an accurate estimate of historical extinction rates for other organisms that do not preserve as well as vertebrates, but scientists can make a reasonable guess. In 1974 the extinction rate for all species was estimated to be approximately 100 per year.[1] Because mammals and birds combined comprise less than half of one percent of all living species, this estimate is probably conservative. Norman Myers projected that by the end of this century, species will be vanishing at the rate of 100 per day, due largely to the destruction of tropical rain forests.[2] On a human time scale, these rates may seem slow. On a geological time scale, however, the wheel is spinning at a blurring rate, and the disappearance of species amounts to a virtually instantaneous mass extinction.

The enormous variety of life is regarded by many as a sort of sideshow: fascinating, but dispensable, should it stand as an obstacle to human interests. This shortsighted view is finally beginning to exact its toll. Animals and plants that have been of great value to humans are disappearing forever. A piece of America's soul died along with the passenger pigeon, plains buffalo, and American chestnut. We are now quickly losing the whales, elephants, tigers, lions, bears, apes, and rhinoceroses. Imagine a world in which our own children no longer know the fantastic wealth of creatures we associate with the word *zoo*. Imagine summertime with fewer birds than during the winter. The brilliant orioles, tanagers, warblers, and wading birds will be gone because their tropical wintering grounds will no longer exist. But the extinction of any one species is of only passing significance to human society. It is the longer-term cumulative effects that should have people worried.

The long-term effects of mass extinction on human society fall along four basic lines: moral, aesthetic, economic, and ecological.[3] In each of these can be found strong arguments for arresting activities that cause extinctions, but in each case the long-term benefits of species conservation must be weighed against extremely tempting short-term gains.

One consequence of perpetrating a mass extinction is that at least some of us are going to have trouble with our consciences. In human terms, other forms of life have a right to exist, and thus to extinguish them is akin to genocide. Although morals have hardly eliminated regular attempts at human genocide, by a strange twist of fate moral persuasion has often worked against practices that threaten critical habitats and charismatic species, such as songbirds and whales. But the bans on whaling and DDT resulted from conflicts between wildlife and profits. When the conflict is between wildlife and human life, the difference between right and wrong is no longer so clear. Indonesian farmers have been moved en masse into the rain forest as part of a national resettlement plan to ease population pressure. As a result they have been repeatedly harrassed by displaced elephants who stomp on the settlers' crops and sit on their houses.[4] Despite valiant attempts to relocate the elephants, there is little assurance that there will always be room in Southeast Asia for both its people and its magnificent wildlife, of which elephants, tigers, and rhinoceroses are only the most conspicuous elements.

A second consequence of mass extinction is that our children might hate us for the world we took away from them. Natural diversity is critical to us aesthetically, helping to satisfy an important human need for sensory and intellectual stimulation. E. O. Wilson calls this hunger for other life biophilia, a basic part of the human psyche.[5] The diversity of life is also the master key to the science of life—as we extinguish species, we snuff out countless lines of fruitful inquiry. But if diversity is a basic need, many people are out of touch with it. Squalid, dirty, lifeless cities make it difficult for us to feel closeness to nature. We do not foster the sense that biological diversity is a thing of value. And yet, it is in this environment that most of humanity may be finding itself in the future. In the noisy cities the aesthetic arguments for saving species fall on deaf ears.

The extinction will have economic impacts of fantastic proportions but of a sort that only a broker in futures can appreciate. This is most apparent in tropical rain forests. The sustainable yield from a rain forest through lumbering or shifting agriculture is low, so tropical industries have tended toward a "get in, get out" modus operandi. Although a few interests have gotten rich this way, new technologies have made this way of doing business increasingly stupid and wasteful. The major commodity in the world's rain forests lies dormant and largely unrecognized: It is information. Sequestered in the countless plants and insects of the forest is a vast chemical arsenal. To the organisms these chemicals are defensive and offensive weapons, but to people they offer medicines and chemical conveniences of great potential.[6] The importance of this chemical warehouse lies not only in the substances themselves but also in the fact that for each there already exists a genetically coded blueprint. It is such a blueprint, extracted and placed in bacteria or yeast cells, that can allow us to produce huge quantities of any desired substance. The genes of rain forest organisms also contain the secrets needed to create animals and crop plants that can live effectively under moist tropical conditions. Just as the synthetic chemical industry has made oil too precious a resource to lavish on Sunday drivers, so has biotechnology made natural diversity too valuable to let it be wasted on narrow-minded, one-shot business ventures. This is one resource that can be protected only by keeping it absolutely intact.

The fourth argument for preserving biological diversity is the simplest: Our lives depend on it. We are part of a common fabric of life. Our survival is dependent on the integrity of this fabric, for the loss of a few critical threads could lead to a quick unraveling of the whole. We know that there have been previous mass extinctions, through which some life survived. As for our own chances of surviving this mass extinction, there can be no promises. If the Grim Reaper plays any favorites at all, then it would seem to be a special fondness for striking down dominant organisms in their prime. David Jablonski examines the fates of rudist clams, mammallike reptiles, dinosaurs, and a host of other scintillating but doomed creatures in his essay. Humans are now the dominant creatures, at least in terms of their influence. So, lest history bear false witness and barring some serious conservation efforts on our part, this mass extinction could well be the last one that we will ever know about.

What Is It That We Are Trying to Save?

There is now a large community of people from all parts of the world who share a common interest in the preservation of biological diversity. Although kindred in spirit, they do not all share the same assumptions about what they should be doing. Three major stumbling blocks are (1) differing views of what biological diversity and its preservation actually mean, (2) differing thoughts on the seriousness of the threats facing biological diversity, and (3) differing notions as to reasonable goals for the future. People are certainly entitled to different opinions, and society can benefit from a diversity of views as well as a diversity of species. Still, much confusion arises from problems that have more to do with lack of knowledge than lack of philosophical agreement.

In the conservation literature, preservation of biological diversity is sometimes assumed to be synonymous with the preservation of species. Unfortunately, the meaning of the word "species" is not always clear. Biologists use it in two senses. The first, the taxonomic species, is the name by which people call a particular form of life. In the second sense of the word, the taxonomic notion has been modified by an appreciation of evolution. Evolutionary relationships can lead to species groupings that are quite different from those suggested by superficial similarities or differences among taxa. Furthermore, a single species can be composed of several visibly different populations. So, although humans generally prefer things neat and clean, evolution is a bit of a mess. The second species concept has made a good deal of trouble for the first one.

The naming business, or taxonomy, is a human invention designed to help people cope with a world of bewildering complexity. Taxonomists group like forms according to their similarities and differences and then pronounce individuals that share a great many characteristics to be members of the same species. Similar species are grouped into a common genus; similar genera into families; similar families into orders, then classes, phyla, and kingdoms in an expanding hierarchy. Without taxonomic distinctions the multifariousness of life could not be appreciated or communicated, or even bought and sold.

Unlike taxonomy, the evolution of species is a natural phenomenon and not a human filing device. Toward the end of the last century, after scientists had already named and cataloged a great many living forms, Charles Darwin and Alfred Wallace began to wonder if there might be a natural mechanism that could account for the diversity of species. One telling observation was that the members of a species all bear close resemblance to each other because they breed among themselves and are unable or unwilling to breed with other dissimilar organisms. A natural species was thus a group of like individuals that maintained their genetic integrity through a series of reproductive isolating mechanisms. Darwin and Wallace realized that this genetic homeostasis is what makes it possible for a species to preserve advantageous traits that develop through natural selection. Thus natural selection both creates and maintains life's diversity.

In order for similar species to coexist without losing their integrity, they must have effective reproductive isolating mechanisms. Species-specific marks and behaviors, such as ritual mating dances, play important roles in ensuring that species find members of their own kind for procreation. This distinctness is maintained, however, only so long as it is favored by natural selection. If a hybrid is accidentally produced, it will usually be a misfit and less well adjusted to its world than purebred members of either parent species. Thus the occasional hybrid individual will generally die or fail to reproduce, and the parent species will maintain their differences. If selection against hybrids is relaxed, the species will mingle, and after a while they can lose their distinctness.

Darwin appreciated that one kind of species can give rise to another, or many others, but he never did quite fathom how this happens. Years later, with the advent of genetics, people finally began to understand how species begat species, and with this breakthrough taxonomy begat the new science of evolutionary systematics. Scientists continued their business of naming newly recognized species, but now they faced the more challenging business of inferring evolutionary relationships. They also began to realize, to their chagrin, that because evolution is an ongoing process, species in nature are simply not as clear-cut as taxonomists would like to have them. This can make conservation of species difficult, for within each species there can be many distinct varieties.

Obviously, different-looking creatures encountered side by side in nature are not necessarily different species. Caterpillars and moths represent different life stages of the same species. A towering cedar and a forest of miniature bonsai nurtured from its cuttings represent not only the same species but also identical genetic material cultured under different conditions. Many butterflies come in a range of genetically distinct varieties, suited to different times of year or ways of life, but all are part of one species. In addition to wild varieties of a species shaped by natural selection, there can be thousands of genetic varieties created and maintained through artificial selection. So, in its broadest context, conservation of biological diversity means safeguarding both species and varieties within species. To an agricultural scientist or a farmer, the preservation of the many varieties of any one crop plant is at least as important as the conservation of wild species.

Some lineages have many species that are virtually indistinguishable to us, and yet they are clearly able to tell each other apart. Other lineages are composed of relatively few species, but these differ greatly from one another in their appearance. Oak trees offer the worst of all possible worlds. They look very different to us yet hybridize freely. The hybrids survive and even abound in environments that do not clearly favor one or the other parent species.

As a college student I made part of my living as a plant geographer, identifying species of oak trees at 40 to 60 miles an hour while driving along the highways and backroads of Maryland. During this assignment we realized that a large percentage of the trees could not be assigned to species. Dry-country oaks lived on the tops of hills and wetlands oaks in the bottoms, but between the two, where most of the roads happened to be, were many apparent hybrids. Gray's *Manual of Botany* lists seventy-four "kinds" of oak trees (*Quercus*) in eastern North America: twenty-seven species and forty-seven hybrids.[7] For botanists it sometimes seems easier to reject the whole notion of pigeonholes. Verne Grant, who has contributed a great deal to our present understanding of plant evolution, has described in detail the myriad ways in which plants fail to conform to a traditional species concept.[8]

Such irreverence for evolutionary theory is not limited to plants.[9] The penguin's formal attire disguises a great contempt for taxonomists. Several penguin lineages have spread to disparate areas and developed distinct local populations.[10] Nobody seems to be able to decide whether they should be called species. Believe it or not, such a thing can really matter. Recently, the New England Aquarium began to assemble a colony of rockhopper penguins, *Eudyptes chrysocome* (figure 1.1). Rockhoppers are the most widely distributed of all penguins, and as things turned out we wound up getting specimens from two different localities. Our first batch came from an island off South Africa, while the next batch came from the Falklands, near South America. After spending awhile with our South African guests, the first glimpse of the South American penguins inspired a feeling of mild culture shock—they clearly were not the same birds as the ones we had been coddling for several weeks. They looked, acted, and sounded different and, as time wore on, proved to have rather different temperaments. All penguins have somewhat bad temperaments—it is part of what makes them so endearing to humans—but our second batch was just a bit worse than usual. Were they indeed the same species?

a

Figure 1.1
Two rockhopper subspecies. (*a*) From subantarctic islands near South Africa. (*b*) From South America. Photographs by Les Kaufman.

b

Figure 1.1 (continued)

Reference to the literature on crested penguins reveals that the rock-hopper is, taxonomically, a quiet corner of an otherwise noisy battlefield. Penguinologists have yet to agree on how many other species of crested penguins there are besides the rockhopper—responsible estimates range from two to five. There is even less agreement on the total number of penguin species in the world: there may be from eleven to eighteen.[11] Most estimates average between sixteen and eighteen, but a rabid "splitter" could bring the total to twenty by raising two well-differentiated subspecies (southern gentoo, southern little blue) to full species status. Given one zealous afficionado and a pair of calipers, the rockhopper itself might awake one morning as at least two species.[12] Although penguins themselves express little concern for how many species they might comprise, the existence of each of these forms is of value and interest to biologists. The expert eye can distinguish at least twenty-one living forms, and, species or no, the extinction of any of these would eliminate one more chapter in a fascinating history.

Scientific curiosity is hardly the sole justification for trying to preserve all distinct forms of life, whether they are species or not. First, our notion of what a species is, or isn't, is largely an artifact of human bias. It is difficult to prove that an organism is a member of a distinct species, even though it looks and acts different from anything else around. Preserving it is the safe thing to do. Second, much of life's diversity is below the species level, especially our priceless store of livestock strains and fruit, vegetable, and grain cultivars. We obviously do not want to lose these, but we forget that natural subspecific variation also represents a library of special characteristics that could someday prove useful.

There is another purely strategic rationale for promoting conservation below the species level. Widely distributed species are often a composite of several geographically isolated populations, each a bit different from the others in appearance and behavior. In the games of environmental chess, an odd little pawn of a population can hold the key to a major victory. Because of the importance of the Endangered Species Act and similar laws in other countries, the future of an entire habitat and all the species in it can depend on the existence of one organism eligible for legal protection. Many a small, embattled wilderness has been saved by the discovery of one or two species of local concern. The eastern populations of seaside alder, *Alnus maritima,* have figure prominently in efforts to preserve the gorgeous tidelands of Maryland and Delaware. (Elsewhere this species is restricted to the Red River of Oklahoma.)[13] Most herpetologists do not regard the Plymouth red-bellied turtle, *Pseudemys rubriventris bangsi* (figure 1.2) as fully distinct from its common Mid-Atlantic cousins.[14] Yet because it is one of the few New England animals now on the United States list of endangered species, it is a local celebrity and the focus of a rehabilitation program likely to save both the turtle and a chunk of the Massachusetts pine barrens.

Regional variants and locally endangered populations are more than levers for protecting habitats; along with green strips and public victory gardens they are keys to the amelioration of urban people's sense of isolation from the natural world. All people yearn to be special, or a part of something that is special. This need is a source of human potential still untapped by conservationists. Everybody has heard of the Siberian tiger, but only recently has the Willow Pond stickleback become an object of pride for the people of Jamaica Plain, Boston. These diminutive golden fish are an unusual local population of one of the most widespread vertebrates in the Northern Hemisphere, the common three-spined stickleback, *Gasterosteus aculeatus.* The Willow Pond stickleback is of some interest to educators and evolutionary scientists, but far more remarkable is the tiny pond it inhabits.[15] Here, in the middle of a filthy city park, is an oasis of cold, crystal-clear, spring-fed water. When the state became aware of the fish's existence, the pond was placed (at least on paper) under protection. Gradually, more

Figure 1.2
A Plymouth red-bellied turtle. Photograph by Les Kaufman.

people are thinking of the park as a small but special part of their neighborhood, and there is hope that it might be better taken care of in the future.

Programs such as the red-bellied turtle rehabilitation do a great deal to raise environmental awareness, but it can be at the expense of time and one that might otherwise be given to more severely threatened forms at the species or genus level. For zoos and aquaria there is a delicate balance to be set between resources allocated to public education and those reserved for direct aid to the neediest species. In his chapter on the Species Survival Plan, Tom Foose makes it depressingly obvious that there is neither the money nor the space to preserve more than a handful of species in the world's zoos and aquaria; one representative form of a species, or species group, is about all we will ever be able to handle in captivity. Ideally, this should be the form with the best chance of being successfully returned to the wild in the future. Unfortunately, those of us who work in zoos and aquaria know that such pragmatic concerns will most likely yield to charisma as the main criterion for choosing species for the zoo ark. By these standards both Plymouth red-bellied turtles and slightly odd sticklebacks are liable to get kicked off the list early. As for crested penguins, I am confident that the rockhopper will win in any contest, for of all the penguins it bears the greatest resemblance to a rock star. For rockhopper penguins, at least, the future is not all that grim.

How Bad Is This Extinction?

One of the principal obstacles facing conservationists is a lack of public awareness of the magnitude of the extinction taking place today. By the most pessimistic accounts, the current mass extinction is more severe than the one that wiped out the dinosaurs 65 million years ago. More optimistic accounts note that we do not know how bad the problem really is because there are no comprehensive, up-to-date data on extinction rates, especially on organisms yet to be discovered.[16] Some have gone so far as to interpret the lack of data to mean that the problem has been greatly blown out of proportion, but most scientists would reject this "know-nothing" position.

Obviously if there is to be substantive movement on this issue, a great many people have to be convinced that there is a problem. The lack of adequate data is a reality. But if we wait until there are empirical data to back our pleas, we will clearly have waited too long—the mass extinction will already be far advanced. Thus the crucial premise that we are presently experiencing a period of mass extinction must rest on inference. How solid is this inference? How strong are its supporting arguments?

Three major claims are at issue: that we are experiencing a mass extinction, that this extinction is caused by humans, and that this extinction demands immediate attention as one of the most serious problems facing the world today.

There is no doubt that a mass extinction is occurring, even though most of the evidence is inferential rather than direct. Even the historical extinction rate of one bird or mammal species per year since 1900 qualifies this as a mass extinction. If you are still not convinced, read Jablonski's review of this extinction as compared with past ones and Williams and Nowak's list of extinctions at both the species and subspecies level in North America. The original discussions of this issue in the popular literature are those of Vinzenz Ziswiler and David Ehrenfeld, followed by the more sensational books of Norman Myers and of Paul and Anne Ehrlich.[17] It is frequently noted that news of the extinction crisis has been brought to us by the same people who brought us the population crisis (now seemingly dormant) and the nuclear winter (a hypothesis which, if ever tested, will render this book superfluous). To some people, these Cassandras, as they call themselves, are professional doomsayers, intellectual terrorists who should not be encouraged, supported, or believed. What seems to have escaped such doubters, however, is that the whole point of being a doomsayer is to agitate the world into proving you wrong or into doing something about it if you are right.

In any event, when it comes to the extinction issue, Myers and the Ehrlichs did more than review historical extinction rates. They projected such rates through the year 2000 and concluded that they would climb to 40,000 species per year—40 million times faster than the rate of extinction of the dinosaurs. This project is based largely on certain assumptions about the rate at which tropical rain forest is being destroyed around the world, as it is in the rain forest that much of the world's diversity is sequestered.[18] According to one critic, Rodger Sedjo, a forestry economist with Resources for the Future, Myers's calculations were based on early estimates of deforestation rates that were highly inflated.[19] More recent data suggest that actual rates are less than half those first projected.[20] Furthermore, Sedjo disagrees with Myers's assumptions that the number of species lost is necessarily directly proportional to the area of rain forest destroyed per year and that a large proportion of what is being destroyed is primary rain forest.[21] In a popular article, John Tierney interpreted Sedjo's results to mean that the entire mass extinction problem has been trumped up.[22] But the new data on tropical forestation do nothing to alter the bleak scenario because a factor-of-two change in projected rates of extinction means nothing when orders of magnitude are at issue. Moreover, much of the primary rain forest land now being cut will not regenerate as rain forest.

Worse still, the relation of species lost to area of habitat destroyed could actually be higher, not lower, than a one-to-one ratio because many of the world's high-diversity habitats have been carved into little pockets where unique species are concentrated. Foose's hairy rhinoceros is limited to a tiny peninsula in the southwest corner of Java. The golden lion tamarin holds court in a narrow strip of remnant coastal rain forest in Brazil. The rain forest of northern Australia, with many endemic forms, has been reduced to a strand of small parks strung along the top of a mountain range and a few lowland areas cut up by land speculators.

A body of ecological theory, called island biogeography, predicts that extinction rates increase disproportionately when continuous habitats are transformed into archipelagoes of small biological islands.[23] Thomas Lovejoy and his associates are now examining the magnitude of this effect in the Amazon by comparing the ability of variously sized remnant patches of Amazonian rain forest to maintain their original flora and fauna (figure 1.3).[24] It is already apparent that some of the most important biological preserves in the world are too small and too highly disturbed to hang on to their treasures: Parched Everglades and poached African game parks come immediately to mind. Some of the most interesting faunas developed their diversity within small areas that are now highly vulnerable. All such cases are, like the famous Galápagos Islands, priceless evolutionary laboratories. One example is the Hawaiian fauna, whose demise is discussed by Williams and Nowak. Lakes Tanganyika, Malawi, and Victoria in East Africa, Lake Lanao in the Philippines, Lake Baikal in Russia, and Lake Titicaca in the Peruvian Andes have all mothered astounding radiations of freshwater life, wild profusions of evolution that make the Galápagos finches look pale.[25] All are either severely threatened or already disrupted.[26]

The Ehrlichs may well have overstated their case. Perhaps Myers stretched his imagination a bit, or his calculator. But the only allowable

Figure 1.3
Forest patches in the Amazon. Photograph by R. Bierregaard, World Wildlife Fund.

interpretation of the data now at hand is that the world is indeed experiencing a mass extinction easily on a par with those of the geological past.

Mass extinction is not the sort of event we would expect to slip by unnoticed, but this one seems to be doing just that. On our own time scale, the extinctions occurring today seem scattered over many years and many miles. They do not seem so bad. Partly, however, people may be denying facts because they do not want or know how to cope with them.

Stephen Jay Gould, of Harvard University, has made a career out of demonstrating the broad chasms between what people want to believe and the truth. He maintains that people most want to see the world in terms of slow, gradual changes, whereas most of the characteristics of the world have been shaped by short-lived events that are, in retrospect, catastrophic. Gould first faced this kind of opposition when he questioned a traditional view that evolution takes place gradually and continuously, with species gracefully branching from the mother stem, as in a willow tree. With his associate Niles Eldredge, Gould countered that most evolution takes place during the geologically brief, tumultuous birth of a species.[27] Once born, a species remains relatively constant in character until it goes extinct. This notion of evolution by "punctuated equilibrium" has gained wide acceptance but only after a struggle. In a recent speech Gould postulated that the same facet of human nature that inspired people to reject the notion that species can evolve quickly and then virtually cease evolving, now has them denying that species can disappear even more quickly than they evolve, never to return.[28] The notion is, after all, rather disturbing. If other species can disappear so quickly, regardless of their strengths and weaknesses, so can we.

To a nonscientist it might seem that extinction, even mass extinction, is a natural process and can be the result only of natural causes. If we are having a mass extinction, then too bad; we must have been born into this universe at an inopportune time for professional biologists, not to mention the rest of us. Yet, as explained by David Jablonski in his essay, the timing of this mass extinction is way out of line with previous ones. Past mass extinctions have corresponded to about a 26 million year cycle. This one was not due for another 13 million years.

Human activity is having an obvious impact on the face of the globe. In the more developed nations, habitat destruction has gone about with such industry that these societies come to resemble gigantic anthills. Verdant hills have deteriorated into vast pit and strip mines. Coastlines alternate between megalopolis and motels. River valleys everywhere have been dammed to yield spreading reservoirs for hydroelectricity, drinking water, and recreation. Agricultural, industrial, and residential real estate developments have pushed the wilderness back into vest-pocket parklands. Each holiday, waves of workers and high-pressure technocrats seek moments of peace in these trampled sanctuaries. Most of the lands and waters have been laid to waste at least once before. Development really means redevelopment, and there is at least as much need for restoration as for conservation.

Now the governments of the less developed nations, nearly all of which are situated in the tropics, feel that the same economic prosperity and material wealth are due to them. They are ready to follow the path laid down by the developed world, and the world powers have been all too ready to help them along the way, taking most of the riches for themselves in the process. As a result, vast reaches of the tropical America, Asia, and Africa are now being drowned for hydroelectricity, logged for hardwoods and pulp, cut clear for cattle, and mined for ore. Where giant military machines cratered and scorched the ground, starving peoples poach the remaining wildlife and cook it over coals rendered from the few trees still left standing. In Africa, rain forests become deserts; in Arabia, reefs become oil refineries. And yet, despite the sheer magnitude of the destruction, despite monstrously wasteful failures (such as the Jari project in Brazil), mankind's impact in the tropics still seems insignificant against all that there is left to save.[29] Development can be reined in and time bought for more careful planning. The impact of development can be mitigated and the value of Earth's richest communities realized in ample time to save them.

Although industrial development is often painted as the arch-villain, other forms of human impact can be equally damaging and far more insidious because they result from a general degradation of the environment, rather than direct killing and exploitation. If we are to reverse the damage already done in the name of human progress, we must have the ability to distinguish the impact of human disturbance from the influence of nonhuman forces. Intuitively, one might expect such a discrimination to be relatively simple. Developmental psychologists tell us that even first graders understand the limits of their own influence over the environment. On the scale of humanity's relationship with the biosphere, however, we are toddlers. It is rare that we can make a clear distinction between disturbances caused by people and those caused by nature. As it happens, Mother Nature is a disturbed lady.

The study of disturbance in natural communities comprises a major branch in the science of ecology. Experimental ecologists get at the guts of a community by deliberately disturbing it. Often the goal is to remove a dominant species in order to watch remaining inhabitants of the community as they accommodate the new situation. Sessile dominants, such as seaweed or trees, are selectively removed with butane torches and hatchets. More mobile organisms are often "removed" by erecting a cage to deny them access to a piece of habitat. Miles of chicken wire have gone to caging competitors and predators either in or out of experimental areas. The consequences of these experiments sometimes have a familiar air about them, for regenerating patches in somebody's experimental community often resemble patches that occur naturally right next door. And that is exactly the point. Natural communities frequently owe their character to regular disturbances. Predators help maintain a high diversity of herbivores by preventing any one from monopolizing space and food. Grazers promote high diversity among coexisting plants and invertebrates in the same way. In more familiar terms, a lawnmower is a grazer, and frequent mowing can hold the crabgrass at bay so that at least a few strands of Kentucky Blue are able to coexist with it.

Physical disturbance can also promote diversity in a community. Forests are in a constant patchwork of succession, each patch in a different stage of regeneration from the last fire or severe windstorm. Recently opened patches are occupied by so-called fugitive species that rush in and complete their furious, short lives before yielding to slower-growing, longer-lived species later on. Thus quick-growing, light-colored aspen coexist among darker spruce and fir to form a familiar mountainlands' tapestry. But although natural disturbances help to maintain diversity, disturbance that is caused or aggravated by humans frequently lowers diversity and threatens the welfare of species. The plot quickly thickens when a careful search through the dead and dying timber of forest clearings fails to reveal any sign of juvenile trees exploding up into the newly opened space.

In 1976, Douglas Sprugel first described what he called fir waves in northern forests.[30] Sprugel observed that, when tree fall or insect attack opens up a rent in the fir canopy on a mountainside, the patch of dead trees, over years, seems to roll up the side of the mountain as a wave of death and regeneration (figure 1.4), like the electric sparks ascending a Jacob's ladder in old science-fiction movies. He postulated that the trees on the uphill side of an open patch are the most exposed to the elements and thus tend to die, particularly in the wintertime. Meanwhile, trees on the downhill side of the patch regenerate quickly. The result is new trees filling in from below as trees along the leading edge of the patch continue to die, causing the ribbon of dead trees to sweep up the mountain. I remember my own excitement at first noticing these disturbance waves in

Figure 1.4
Fir waves. Photograph by Douglas Sprugel.

the mountains of Nova Scotia and Newfoundland and later on the tops of the Great Smokies. There are even signs explaining that large areas of dead trees on Mount Mitchell are the result of natural disturbances, such as fir waves and outbreaks of insect pests. But lately the death of trees at high altitudes has not been following its expected course. Now park interpreters bear a different message to their visitors. There is suspicion that acid precipitation and other forms of atmospheric and soil pollution are major contributing factors to this death, and the public is educated not only in the cyclical ways of nature but also in the depredations of mankind. Same trees. Roughly the same scenario.

It might seem that species adapted to regular recovery from natural disturbance should also be robust against human disturbance. The truth can be just the opposite. Human impact is in addition to, not a substitution for, those naturally visited on wild populations. This addition of insult to injury can push naturally stressed species and communities beyond their limits of adaptability. Furthermore, the combination of human and natural disturbance can make it virtually impossible to tease apart the powerful swing of nature's pendulum from the seething pressures of humanity. This is particularly apparent in two of the Earth's most sensitive biotopes: the tropical coral reef and the temperate estuary.

For a long time coral reefs (figure 1.5) were thought of as classic examples of stable equilibrium communities. With the advent of scuba gear and long hours of underwater observations, scientists began to appreciate that nearly all coral assemblages experience high rates of biological and physical disturbance. Unappetizing though they may seem, stony corals are consumed by snails, fishes, urchins, starfishes, bristleworms, and even other corals. Corals are wracked by mysterious plagues as well as by slight changes in the physical environment. But strangest of all the hazards a living coral colony must face is its conversion into a hanging garden, a form of piscine agriculture plied by three-inch-long, territorial damselfishes. A damselfish garden is a carpetlike mat of fleshy algae (figure 1.6). The dark mass of algae is conspicuous on a well-populated reef because herbivorous fishes and sea urchins normally keep all algae cropped to a productive but nearly microscopic stubble. Damselfishes keep their gardens going by vigilantly chasing algae-eating creatures from their territories. This helps the algae build up to the point where it attracts tiny shrimp and crabs. The damselfish feed on the algae as well as its animal denizens, and they will even weed the algal garden of undesirable species.[31]

It is common for human disturbance to a community to be mislabeled a natural event, but damselfish gardens are one case in which a natural

Figure 1.5
The harmony of a coral reef. Photograph by Les Kaufman.

Figure 1.6
Three-spot damselfish tending an algal garden on a coral reef. Photograph by Chris Newbert.

disturbance was mistakenly blamed on human fishermen. In the late 1970s it was discovered that certain damselfishes create space for their gardens by actually killing corals (they worry them to death); they then use the dead coral branches as a sort of trellis on which to grow the algae.[32] The life of a territorial damselfish is a cycle of gnash-and-burn agriculture: killing coral, establishing algal gardens, and eventually moving on across the reef and killing more coral. At first this story came as a bit of a surprise to coral reef ecologists, because the patches of disturbed reef were covered with algae. The fish were covering their tracks, so to speak.

    Soon after the discovery that damselfish actually kill coral, a few scientists suggested that damselfish disturbance on coral reefs was an un-natural result of human activity. Damselfishes, it was reasoned, are preyed on by larger fishes that are food for people. These larger fishes have been heavily overfished on the more populated islands for many centuries. Ecologists also work on populous islands, for that is where marine labs, good beer, and chicken wire for caging experiments can all be found—hence the hypothesis that damselfish infestations are a recent development, triggered by overfishing of damselfish predators, limited to heavily populated islands, and dutifully observed there by a somewhat provincial crowd of marine biologists. Some workers even expressed their concern that the reef was in danger, soon to be scoured by the damselfish scourge. As things turned out, algal gardening damselfishes proved to be an abundant and integral part of reef communities all over the world, regardless of fishing pressure. One can even find fossils of algal gardens from more than a hundred thousand years ago, long before overfishing could have been a factor.[33] This disturbance, at least, is a natural one and one important determinant of coral community structure.

For all its ability to generate disturbances, the human species is relatively ineffectual at controlling them. The story of the crown-of-thorns starfish (*Acanthaster planci*) is the stuff of which great disaster movies are made. In the early 1970s, this rare and most unpleasant animal began turning up in unusually high numbers around Green Island, on the Australian Great Barrier Reef (figure 1.7).[34] A predator on live coral, the starfish soon proved up to its legendary appetite by almost totally eliminating the coral cover that once surrounded this important tourist resort. The initial starfish hoard grew quickly and rolled like a spiny juggernaut across the northern Great Barrier Reef. Through popularized accounts of his work on *Acanthaster*, Robert Endean became a dark prophet of doom. He predicted that in the near future the starfish would destroy great portions of this greatest of living

Figure 1.7
A crown-of-thorns starfish infestation in Australia. Photograph courtesy of the Great Barrier Reef Marine Park Authority and *Oceanus* magazine.

coral reefs, the most diverse animal community on Earth. Endean pinned the blame for this unprecedented disaster on human activity—perhaps overzealous shell collectors destroying one of the starfish's main predators, the Triton's trumpet shell (*Charonia tritonis*) perhaps some other less obvious form of human mischief.[35] Scientists were shipped out to scan Pacific reefs for wandering swarms of voracious crown-of-thorns. Teams would descend like Valkyries on the advancing armies to administer to each starfish its final sentence: capital punishment by lethal injection.

Most of these attempts at mass annihilation were inadequate—the starfish just kept coming. Until suddenly and mysteriously they stopped coming. Having consumed all the living coral they could find, the adults marched off to unknown quarters. Research during the subsequent decade indicated that crown-of-thorns starfish have erupted periodically on the Great Barrier Reef, perhaps once every four hundred years or so.[36] Certainly this suggests that at least some previous infestations had little to do with human interference. Other research ties good years for crown-of-thorns recruitment to heavy rains and the associated low salinity and high productivity, a situation that could favor the differential survival of crown-of-thorns larvae.[37] Research money for work on *Acanthaster* disappeared along with the starfish, and the degree of human involvement in the crown-of-thorns case was never firmly established. Even as some of the best marine scientists are trying to pry their way past hedgerows of conflicting data from previous infestations, the starfish have returned.

In 1984, a small troop set sail from Cairns, Australia, on the work-boat *Hero* for three weeks of research on coral reef fishes. Our initial intention was to conduct a comparison of a few reefs on the middle and outer portions of the continental shelf. All did not go as planned. The expedition turned into a wandering journey north through the Barrier complex in search of reefs that still had decent live coral cover. The crown-of-thorns had marched again, breaking trail through an estimated 86 percent of the reefs of the central barrier. Some of the most pitifully devastated reefs, showing less than 10 percent live coral cover, were known to have been fully regenerated since the previous starfish infestation, a period of only eight or nine years.[38] So, biological lightning does strike twice in the same place, but this is only possible because of the reef's surprising regenerative capacity.

The secret of this resiliency was visible in the eerie landscapes of reefs killed by *Acanthaster* during the previous couple of years. Big coral colonies stood like upright mummies, the old, dead coral skeletons encased in a thick jacket of encrusting calcareous algae. On some reefs, though, these limestone coffins were studded with vigorous young coral colonies. Herbivores seemed to play an important role in this rapid regeneration. Schools of grazing fishes, such as the giant bumphead parrotfish, moved over the reef, scraping bits of algae from the surface. This grazing pressure kept a canopy of fleshy or filamentous algae from forming on the limestone surface of the reef, making it easier for baby corals to find space in which to establish themselves. Ironically, the fringing reefs of Green Island itself have yet to show strong signs of recovery. This may have to do with any number of factors: for example, the island's proximity to land and concomitant problems with turbidity, freshwater runoff, lower densities of giant parrotfishes, and greater impact by tourists and fishermen. We may know soon because the Australian government has again committed itself to comprehending this morbid starfish. A major *Acanthaster* program is underway at the Australian Institute of Marine Sciences, and the jury may soon be in on mankind's hand in the Australian disaster.

Even if human activities are not the immediate cause of a catastrophic disturbance, they can lead to other misadventures that seriously compromise a community's ability to return to its original state. Recent observations in the Caribbean bear this out in a tale that is in some ways even more gruesome than that of the crown-of-thorns. In this case, catastrophe rode on the wings of monster storms, not monstrous starfish, and the path to recovery was strewn with obstacles, both human and natural in origin.

The submarine topography of Jamaica's north coast is a spectacular series of cliffs and abyssal canyons set among broad, gently rolling terraces. Perched on the terraces are coral reefs that were once among the most beautiful in the world. It was here in Jamaica that modern coral reef biology first took shape, to the strains of reggae, under the energetic leadership of Thomas Goreau. Though he traveled and dived all about the island, Goreau centered his work in two locations: the old pirate town of Port Royal and Columbus's "dry harbour," Discovery Bay. By 1980, more than a decade after Goreau's untimely death, the Discovery Bay Marine Laboratory was a major center for graduate studies in marine biology and is a gateway to the best-known tropical sea bottom in the world.

On August 9, 1980, Hurricane Allen plowed through the straits between Jamaica and Cuba. One of the most powerful storms of this century, Allen beamed a rain of titanic waves at Jamaica's north coast. For twelve hours, wave after wave crested up to 13 meters over the shallow reef. By the following morning, the coral reef of Discovery Bay (figure 1.8) was pulverized to depths of nearly 20 meters, and the scientists who had worked at the laboratory awoke to a new outlook on life. For years we had been viewing the reef as an equilibrium system, held in a system of checks and balances by interactions among all the component species. Then in one

a

b

Figure 1.8
Staghorn coral at Discovery Bay, Jamaica, before and after Hurricane Allen. The numbered tag indicates a study site. Photographs by Les Kaufman (*a*) and James Porter (*b*).

evening virtually everything was stripped from the surface of the shallow reef, including our experiments. Dazed fishes milled about homelessly. Coral predators concentrated on the remaining scraps of living coral, like elephants about diminishing waterholes. Herbivorous fishes, already undersized and severely depleted by fishermen, were even scarcer than usual in the early days and weeks after the hurricane. Sea urchins were also scarce for the first few weeks, and consequently algae began to take over the reef surface, coating it with flowing masses of pink, brown, and green. Storms with the fury of Hurricane Allen pass a given spot in the Caribbean about once every hundred years. Hardly an equilibrium system. Instead, as we realize now, every reef is a reef in the process of regenerating from a century hurricane.[39]

By late 1983, two years after Hurricane Allen, the Discovery Bay reef was showing healthy signs of recovery. The herbivorous urchins were back, and much of the weedy algae was gone, leaving behind many clean, grazed surfaces coated with hard coralline algae. New growths of staghorn and elkhorn corals, the delicate, plantlike colonies that lay the framework of a Caribbean terrace reef, were sprouting everywhere. The branches of these new and recovering colonies were growing outward at the furious pace, for corals, of up to 20 centimeters per year. It was like a New England spring on the Jamaican reef. The biologists who had come to love this corner of the ocean as their private garden were at last beginning to feel optimistic about its future. This optimism was to be short-lived, however, for at the height of our confidence in these signs of recovery the reef was struck by a new and bizarre disaster. In just a few short weeks an unknown agent almost entirely wiped out the long-spined Caribbean sea urchin, *Diadema antillarum*.[40]

The disappearance of *Diadema* was at first welcome news to tourists, marine scientists, and graduate students. Our memory of this creature during its heyday in Jamaica is one of pain-tinged nostalgia, for with its evil, violet-black spines abristle, a single long-spine urchin can translate one errant step into days of agony. A few weeks after the die-off, the news was not so happy. With so few grazers about, the entire shallow reef was soon covered with a thick mat of fleshy algae (figure 1.9), and the gains made by corals in the previous three years quickly evaporated. Once-vigorous young colonies were discovered dead and covered with algae, all hope lost for any quick regeneration. This urchin population bust could well have an impact in the Caribbean as serious as the crown-of-thorns populations booms in the Pacific.

Hurricane Allen and the mysterious death of the urchins were both short-lived, natural events, but blame for the severity of their aftermaths

Figure 1.9
Coral reef after a sea urchin die-off. Photograph by Chris Newbert.

could lie squarely in mankind's own court. The removal of large predators from a reef probably does not make or break the territorial damselfishes, but direct fishing of herbivores has definitely depressed their populations. The Jamaican reef is heavily overfished, and the schools of large parrotfishes typical of undisturbed reefs are conspicuously absent. Had the original standing crop of herbivorous fishes not been so greatly reduced over the years, the sudden loss of the urchins might not have had such disastrous consequences. In this way a chronic, long-term human impact could have rendered the community unusually fragile in the face of sudden and un-predictable events, such as the urchin plague. *Diadema* are gradually return-ing to the reefs of Jamaica, but five years after Hurricane Allen, we are still waiting for the coral to regenerate significantly.

Impact on Coastal Estuaries

Whereas coral reefs provide at least an illusion of constancy, estuaries are worlds of intense and constant change. Even their definition bespeaks violence, for estuaries are drowned river valleys. They draw their pulse from the tides, but it is offbeat. Complex shorelines and the mixing of fresh and salt waters can make reference to tide tables a meaningless ritual. In a system with such flux, it is especially difficult to distinguish natural from human disturbances. The two together can be lethal to a species, even for organisms adapted to great environmental extremes. This is nowhere so clear as in the largest of the Mid-Atlantic estuaries, the vast reaches of marshland and tidal creek celebrated in James Michener's novel *Chesapeake*.

The lush harvest of creatures drawn from the Chesapeake's productive waters includes two fishes of special economic concern: the menhaden (*Brevoortia tyrannus*) and striped bass (*Morone saxatilis*). These two species are a study in contrast. The striped bass is a predator, the menhaden its prey. The striped bass is historically among the anglers' most sought after quarry, game to catch and delicious to eat. The menhaden is a virtually inedible plankton-feeder that will not take a hook, but its rich oils render it the most economically valuable species of finfish in the western Atlantic. Striped bass ascend estuarine tributaries and rivers to lay their eggs. Menhaden spawn far out at sea. But from the moment of spawning, each day brings the larvae of these two species closer to their common fate.

Menhaden and striped bass are but two out of tens of fish and invertebrate species that use coastal Atlantic and Gulf marshes as nurseries. It is odd that in order to bring their larvae to this one place, the parents of this great confluence of larvae should expend so much energy in migrations to opposite corners of the sea. Winds and currents drive the young of menhaden from the open ocean to the mouths of estuaries, where they dive to the bottom and catch up-estuarine currents to the headwaters. Meanwhile, the young stripers hatch from their eggs and tumble downstream with the rains of late spring. The two worlds meet between April and June along the meandering, grass-bordered tidal creeks of the Mid-Atlantic estuaries, in the shadows of the great intake and effluent pipes of electric power generating stations.

Estuarine power stations (whether nuclear or fossil fuel) are placed far enough upstream to benefit from the reduced corrosion allowed by fresh water but far enough downstream in the estuaries to have enough cooling water to draw through their heat exchangers. These unfortunate logistics, coupled with a huge demand for cooling water, make electric power stations major predators on larval fishes. Untold millions of larval and juvenile fishes are sucked in by the plants' inhalations, then shot through miles of cooling pipes, and finally spewed forth—mashed, embolized, or roasted—in the excurrent stream. The impact of hungry power plants on fish population dynamics has been difficult to assess because the numbers of larvae produced in a given season do not necessarily predict the number of fish that will successfully recruit to the adult population. Efforts have been made to reduce this larval mortality, but much remains as an unavoidable consequence of our energy-rich existence.

Great crops of a few species are typical of the estuary during its growing season, but great kills are also typical, making assessment of the human impact on juvenile fishes especially frustrating. As the summer wears on, the young menhaden grow rapidly and begin to work their way down into the open waters of the estuary. All seems well until early summer, when menhaden suddenly start appearing near the surface, floundering about or swimming crazily in tight circles, forcing their heads above the water in pathetic gasping leaps. On occasion, massive kills pave the surface of the water with dead fish. Some of the kills are casualties of human activity, as when schools of menhaden are lured into warm deoxygenated waters at the base of a power plant and die. Usually, however, people have little to do with the summer menhaden kills. Many of the fishes die of a cluster of diseases, collectively referred to as whirling disease.[41] Others may suffocate in the midst of the great plankton blooms called red tides, and sometimes the bays are slick with oil and scattered shards of menhaden bodies, left behind by massive bluefish attacks. And yet, despite this carnage, the menhaden have been back, year after year, in good numbers, and indications are that, if prudently managed and if no new insults are added to those already provided by nature, the menhaden should return in the future.

Not so for the striped bass. The waning of Atlantic striped bass populations is one of the most celebrated of recent environmental causes, for it has almost certainly been caused by people. The major striped bass nurseries in the Mid-Atlantic are the Chesapeake Bay and the Hudson River, from which spring the coastal populations that range as far north as Nova Scotia. Recent years have seen a precipitous decline in the abundance and size of striped bass, mostly attributed to overfishing. The problem is clearly more than one of exploitation, however, because the adults that have made it upstream to spawn have had little spawning success—a reduced spawning season in certain rivers, abnormal development of eggs and larvae, and outrageously high larval mortalities. So many things have gone wrong for this species that it is hard to know where to place the blame.

In a good year, striped bass place their eggs far upstream before returning to the estuary. The eggs wash downstream until they encounter the first hint of saltiness. Here, in valleys of eelgrass, the young bass feed on plankton and soon transform into the juvenile form. As they grow, the small stripers gradually work their way downstream into water of increasing salinity, until by late summer the yearlings school over sand and gravel beds on the more exposed shorelines.

A number of events have transpired over the past few years that have made life difficult for the Chesapeake striper. Most of these problems were generated by people. Human demands for drinking and irrigation water have caused drastic reductions in freshwater input to the head of the bay, and studies have even been conducted by the Army Corps of Engineers to assess the potential impact of a total elimination of the flow of water into the Chesapeake. The damming of the Susquehanna River at the northern head of the bay greatly reduced access to former breeding areas for stripers and other anadromous fishes, causing a shift to egg laying below the dam, in an area called the Susquehanna Pool. In later years, the center for bass breeding in the upper Chesapeake Bay again shifted to, of all places, the heavily trafficked Chesapeake and Delaware Canal. Other less important breeding areas remain in the Potomac, Choptank, and Nanticoke rivers.

Because striped bass is an anadromous species whose care and maintenance pose no special problems in a fish hatchery, it should respond well to management efforts, such as stocking programs. And indeed it has, but not in the Chesapeake Bay. Striped bass are naturalized in California and until recently were doing well there, especially in the San Francisco Bay area. Striped bass are also a favorite fish for introduction and management in large reservoirs. Lake Marion in South Carolina is home to a famous landlocked fishery. Although this and most other landlocked striper populations are based almost entirely on hatchery stock, it does make one wonder why management efforts have met with so little success in the Chesapeake. Despite concerted attempts, there is a nearly total die-off of hatchery-reared striped bass larvae at the 15- to 30-day stage of development.[42] Among the most frightening explanations yet offered has been the implication that the decline in striped bass is only part of a larger, possibly global decline in anadromous fish stocks due to acid rain. These normally marine animals spend the most vulnerable portions of their lives in fresh or near-fresh water. Many anadromous species are especially sensitive to low pH and sudden shifts in pH, such as the shifts that occur during bursts of acidic spring runoff from snowmelts and early spring showers. They may also be sensitive to industrial wastes that accumulate to high levels around many heavily populated spawning areas. Some of these hazardous materials are at their most toxic to aquatic life when the pH of the water and adjacent soils is low.

The spawning period for striped bass in the Chesapeake has been growing shorter and shorter over the years. At one time it stretched out over two months, beginning in early April and ending in late May. For some years now, the spawning period has been as brief as two weeks. But although a truncated spawning season can make it more difficult for the population to recover, it was not the cause of the initial decline. A reduction in the number of breeding bass was noticed well before the change in breeding season became apparent.

Most intriguing of all the hypotheses for the stripers' decline has been the indictment of a tiny virus called IPN, or infectious pancreatic necrosis.[43] The virus is one of several types of whirling disease, so-called because of the peculiar swimming motion exhibited by its victims. The odd thing about striped bass and IPN-like whirling diseases is that the adults never whirl. In fact, carriers show no sign of illness from the virus, although it is present in their bodies and is passed on to the next generation through the eggs and sperm. It is the striped bass larvae that suffer, with nearly all infected individuals succumbing within fifteen to thirty days of hatching.

Because the IPN virus is a natural disease agent, it might be presumed that the striped bass is in the throes of a natural death, much like that of the *Diadema* urchins. Some suspect that adult Chesapeake bass pick up the virus from their diet of menhaden, notorious victims of both viral and bacterial whirling disease. But then, why wasn't the virus always present in the bass population? Nobody is sure that it has not been. In fact, there is no truly persuasive evidence that IPN, or any similar virus, can be directly implicated in the decline of the striped bass. One thing about fish viral infections we do know, however: they have a special fondness for the executive fish, born into circumstances of high stress and dirty water. Fish viruses tend to break out in fish stocks at high density, such as in a rearing facility or perhaps in an estuarine nursery. Hatchery-reared stock and water from a hatchery can serve as its carriers. The more we extend our intensive management efforts into formerly undisturbed environments, the more we are going to experience problems with exotic pathogens.

So there is no single culprit in the striped bass story. Predation on eggs and larvae by power plants, once a relatively insignificant source of mortality, could now be significant because the future of each year class is held in so few baskets. Pathogens and pollutants are siblings in disaster—susceptibility of young striped bass to viral infections is almost certainly increased by unfavorable or fluctuating temperature, salinity, or pH, as well as all forms of pollution. It is likely that fishing pressure on the adults brought the species to the point where it could no longer meet its losses. But with so many likelihoods, it is hard to find a clear reason why the bass are vanishing or what can bring them back. On the whole, the situation for native Chesapeake Bay striped bass is rather bleak, and we may well be witnessing an aquatic version of the passenger pigeon on its final flight. The irony is that there are still some places for the bass to spawn and ample tidewaters to nurse its young. If this animal suffers extinction, as well it might, it will not owe its demise to anything so simple as overfishing or habitat destruction alone. The striped bass will be the perfect example of a beloved species lost through the insidious thread-by-thread unraveling of its environment.

Shaping the Future

An issue like that of the striped bass can suddenly galvanize a large number of people to the cause of preserving nature. With this problem of awareness out of the way, a new one looms ahead: even conservationists vary in both their motivations and their goals. Conservationists can be politicians, academics, engineers, business people, parents, bureaucrats, outdoorsmen. One conservationist can be watching a bird while another shoots it. Our conceptions of nature vary, as do our cultural biases about what constitutes a friendly, humanized environment, an appealing wild place, or an intolerable wilderness. We can never bring a halt to this mass extinction unless two commonalities are first achieved. The first is a common pride in assuming our place beside, not above, the rest of nature. The second is a common ecological vision to guide us in our forced role as global heads-of-household. It is this latter role that is particularly troublesome.

From the standpoint of species survival, we desperately need to ensure places in the wild for endangered species to hold their own under natural conditions (see Foose's chapter). Most conservationists are aware of this need, but much of the environmentally conscious community is more seriously concerned with the welfare of game and other particularly popular or charismatic species. Because we so often play favorites, community restoration projects often result in either a gigantic suburban garden or a sportsman's paradise rather than any natural Eden. The results are the mélange of alien plants, fishes, birds, and mammals that populate our "conservation lands" and a society in which an expert in wildlife management is often just somebody who knows how to raise exotic pheasants for hunters to shoot.

The sheer extent to which we have moved species about on the globe in hopes of improving things is beyond belief, and the results have been a strange mixture of fortune and folly. Florida is an especially interesting case. Decades ago, Florida lost its native parakeet, but today you can take a walk through downtown Miami and feel as if you are in a neotropical aviary. Roger Tory Peterson's *Field Guide to the Birds of Eastern North America* now provides field marks for fourteen species of parrots! An entire fish fauna has escaped from the tropical fish industry into Florida's canals with no beneficial results (see Williams and Nowak's essay). No, maybe there was one. A friend of mine had to suspend her graduate fieldwork in Nicaragua because of the political turmoil there but soon discovered that the two species of cichlid fishes she was studying had some time ago been introduced into Florida. She moved to Florida to continue her work and was even pleased to discover that her Spanish was still useful.

The rationale by which communities are deliberately modified can be quite opaque. Take, for example, the vigorous attempts at salmon restoration in the American Northeast. Aside from overfishing, the Atlantic salmon has been bested by dams and pollution throughout its native range in the western North Atlantic. Now, despite noble efforts to restore water quality, there are few Atlantic salmon runs left in the United States, although Canada is still faring a bit better. Atlantic salmon is, in the long run, a more desirable candidate for aquaculture and management than any of the Pacific species. This must explain why, in those few spots now suitable for salmon restoration in Massachusetts, it is Pacific coho salmon, not Atlantic salmon, that has often been introduced. Ironically, the Atlantic salmon is one of the largest members of its genus, and because of this, there has been serious thought of introducing the Atlantic species along the Pacific coast of North America. It is pointless to remind people that the Pacific Northwest is already blessed with the richest natural salmonid community in the world. Some claim that the Atlantic salmon is superior to any of the native species there. Perhaps we should arrange a giant airlift to transport all of the Washington coho to Massachusetts and all of the Atlantic salmon to Puget Sound, where they are wanted.

Even when it is obvious that people are endangering the well-being of a community, as in land and shorefront development, it is rarely a simple matter to decide how such an impact should be lessened or reversed. We have been mismanaging species and communities for thousands of years. Everybody knows that we must do a lot better than we have at community planning. The awful truth is that we are just now learning how to go about it. And we will never learn unless something is done quickly to educate a new generation of bold, competent ecologists.

In her book on the disappearance of the world's rain forests, Catherine Caufield echoed Peter Raven's lament over the severe shortage of ecologists, especially ones experienced enough to make management decisions for large tracts of rain forest.[44] The same can be said of coral reefs, estuaries, sea grass beds, or any other large-scale ecosystem. It is not that field biologists are failing to work hard as educators both at home and abroad, as illustrated by Prance's work (chapter 3). But we are constantly wondering what will become of our students. Ecologists somehow never quite "made it" alongside those professionals whose services are viewed as essential to society. Perhaps this is why they have only poorly developed the sorts of applied skills that should be used on a regular basis.

This point may seem a bit self-serving in a paper written by an ecologist, but examine the logic. If you have a medical problem, you go to the doctor. If your high-tech computer goes on the blink, you have a high-tech service representative troubleshoot it. If there is a nasty stain in your furniture upholstery, you let the cleaners take care of it. But what about a messed-up ecosystem? We have yet to produce a school of competent ecological engineers, the U.S. Army Corps notwithstanding. We still do not understand the community fabric that such engineers aspire to fit and cut. Producing such a task force will not be easy because community dynamics are not always governed by simple, general principles. Much of nature is a kaleidoscope of shifting opportunistic links and associations, unpredictable in advance, unexplainable in retrospect. Community ecologists are like Polynesian navigators, tracing their way across thousands of empty miles, led by subtle patterns on the surface of a turbulent sea.

Another much-needed form of education is the conversion to a worldly way of thinking of those people who are unsympathetic toward or unfamiliar with the conservation mentality. The need to preserve biological diversity has become a cause célèbre among the greater community of scientists, naturalists, and conservationists, but the message has not been adequately delivered to most of the general public. Slick science and nature magazines are one effective means, although there is a limited audience here. It is in television that we find the most powerful vehicle for promoting environmental awareness on a large scale.

At present, most of the environmentally sensitive programming appears on public television. Although nature spectaculars attract a large audience, most regular viewers of public television are seasoned converts to the environmental way of thinking. These people should by now be receiving a different, more sophisticated message that the producers are, for the most part, failing to provide. Too often, nature and science shows offer a fare that is predictable and sophomoric. Afficionados of public television have come to expect every natural history film to end with a cacophony of chainsaws, bulldozers, fishing fleets, and wild animal collectors busily dese-

crating nature as the announcer solemnly intones his final, moralistic mono-
logue. The point of all this may well be reasonable and correct, but anybody
still listening has heard it all before. The message is lost. The final five
minutes would have been better devoted to another nifty animal.

The goals to which we should be aspiring are simple: provide incisive
analyses of environmental problems and make concrete suggestions about
things an individual can do to help solve them. Doing this requires a high
level of expertise, creative effort, and production standards that do not come
cheap. When production deadlines and money are both extremely tight, as
they are now in public media, these constraints cut deep into the thorough
research and analysis that are required to maintain high quality in any form
of media production.

As for that other, larger audience of people who are not yet interested
in conservation, a different kind of medium is needed. These people live
in a commercial arena, and they do not read much. If you want to find
them, they will be watching commercial television and feature films. Educators sometimes look disdainfully on these outlets, but they are overlook-
ing their most important audience. Conveying substance through a medium
of entertainment calls for a rare ability to communicate with emotion and
sensationalism, rather than reason. The audience is not stupid, but you do
not have their attention because they do not want to immerse themselves
in deep and troubling thought. To reach them the writer must render
complex, gray issues into clear black and white. It can be done. And it is
worth it.

In 1985 John Boorman released a film called *The Emerald Forest,* based
on a true story about a man building a dam in the Amazon. The man's
little boy wanders off from a family picnic at the edge of the deforested
construction area. The chief of a tribe of friendly natives takes pity on him
for his life "beyond the Edge of the World," and decides to raise him as a
proper, happy member of the Invisible People. The rest of the film is about
the man's ten-year search for his son. It is an adventure film, pure Holly-
wood, with at least its share of violence, barely credible situations, and
gorgeous, naked women who, with all due respect, look more like Hawai-
ians than Amazonians. The theater was full of people whooping and shout-
ing and empathizing during the entire film. But when they left the theater,
everybody was talking about deforestation. As hard as it was to believe the
story itself, it was even harder to believe the effect the film had on the
audience. I left for home shaking my head, wondering if I hadn't taken a
wrong turn somewhere by getting involved in making television nature
films.

Generating interest in the news about mass extinction is important, but people must know what to do with the information once they have it. As a first step, we must come to think of choices in our way of life as a form of voting on global resource issues. Each of us makes decisions about global resources all day long. The average middle-class American delights in both red meat and teakwood furniture but is blissfully unaware that by satisfying these seemingly innocent whims he or she is darkening the future of tropical rain forests.

If the products of biological diversity seem far removed from everyday life in a developed nation and the notion of individual choice an unlikely conservation strategy, just look around. Today people drive smaller, more efficient automobiles. They use water-conserving shower heads and toilets in their homes. They are more careful in their use of electricity and heating fuel. If people can learn to conserve water and energy, they can learn to conserve biological diversity. Values and lifestyles can change so long as there are people willing and farsighted enough to fight for these changes.

For an issue so important, so clearly relevant to the common good of humankind, the constituency of souls devoted to the preservation of biological diversity appears deceptively ragtag and disorganized. The enemy is clear and formidable—it is all those people who would put short-term financial gains before the welfare of our children. But where is the good army that will fight this battle? Believe me, it is there. And I warn any who stand in its way, it is powerful. But it could use a little more organization and a good marketing department.

A case in point is Australia's Greater Daintree Region, an area of spectacular mountains, upland and lowland rain forests, and fringing coral reefs. In 1984 a small anachronistic army of hippies (a.k.a. greenies) emerged from their minibuses and forest retreats to fight on behalf of the Daintree rain forest. The major television networks were there to greet them. The greenies buried and booby-trapped themselves and joined hands across the dirt roads of Far North Queensland to stop the bulldozers, en route to fresh incursions into the Daintree forest. As they must have realized, they were too late. The planned road would not be stopped, and much of the Daintree forest had already been carved into parcels and was the target of heated land speculation by young urbanites to the south. On television and in the papers, the operation looked well meaning but a bit pathetic. The road represented progress and seemed inevitable. Its opponents seemed like a handful of bleeding hearts. The news story, late in the broadcast, came across as a local human interest story, not national news. Nothing could be further from the truth.

The Daintree rain forest is the world's last significant tract of lowland Australian tropical rain forest. It is one of the most interesting and valuable pieces of land on the planet. The Daintree is home to some of the world's most spectacular living creatures, including the fantastic flightless bird called the cassowary, two species of kangaroos that live only in the treetops, and the Cairns Birdwing, a gorgeous gigantic butterfly. The foes of the Daintree forest are not worshipers of the great god Progress but rather a handful of politically astute individuals out for somewhat dubious short-term gains. These include tin mining, logging, cattle grazing, sugarcane growing, and real estate sales. The forest's friends are not limited to a few "alternative lifestylers," for counted among them are some of Australia's most prominent scientists, writers, artists, and private citizens.

Use of the Daintree forest as anything other than a natural preserve simply defies common sense. The Daintree tin mines are not essential for meeting tin production quotas, and, in fact, one major company reputedly lost over a million dollars in 1981 alone.[45] Here, as in other rain forests around the world, the tropical hardwoods cannot be logged without severely altering the forest, and it is doubtful that they can even be considered a renewable resource.[46] Furthermore, rain forests of any kind occupy only one quarter of one percent of the Australian mainland, a land over most of which the traveler aches for the sight of but a single tree.[47] The thought of cutting down a small, priceless forest to add a minuscule fraction of a percent to the world supply of meat and sugar is absolute lunacy. Finally, the real estate value of the land is dependent on the presence of the rain forest. One cannot both cut down the forest to build houses and leave it up to attract buyers. Elimination of the forest also leaves real estate unprotected against the savage rainy season floods typical of clear-cut rain forest lands.

So here is a situation with an obvious right and wrong way about things and a fairly strong and vocal contingent on the side of right. How can there be any contest? Why is it a Greater Daintree National Park that is in question, rather than dusty roads, useless tin mines, parched rangeland, and unlivable subdivisions? The answer is political. The conservation movement in Queensland has yet to organize into a sufficiently powerful political force to effect its ends. But the makings are there. The Australian Conservation Foundation may provide the focus for gathering together the various interested parties. All hopes are on this happening before there is nothing left to protect but cows and sugarcane.

All over the world, there is an army of protagonists fighting for the preservation of biological diversity. But like the heroes and heroines of the

Daintree forest, they do not all know each other yet, and they certainly do not think of themselves as members of a body politic.

It comes as a surprise to many people that private citizens and local conservation groups can actively safeguard species. Along with the zoos battling to keep antelopes and rhinoceroses in this world, there are hundreds of game parks and wealthy cattlemen and not-so-wealthy but devoted individuals who are already playing a significant role in the preservation of endangered mammal species and rare domestic strains. Along with the major botanical gardens of the world are millions of plant fanciers who can and do aid in the preservation of precious plant species and varieties. Hope for the living evolutionary laboratories of the African Great Lakes may well lie with the world's aquarium hobbyists, whose fish tanks may soon harbor the last remnants of the hundreds of endemic fish species now threatened by alien introductions and overfishing.[48]

Citizens and local conservation groups are also safeguarding habitats. Besides the big conservation organizations (for example, Audubon, Sierra Club, National Wildlife Federation, and the World Wildlife Fund), most municipalities have conservation committees of one sort or another, and many government posts in conservation are held by citizen appointees. In Massachusetts and a few other states, for example, citizen volunteers aid the government in setting priorities for the spending of state funds to preserve critical habitats. The funds themselves come largely from a voluntary tax checkoff, whose existence owes much to grass roots lobbying. The idea of a "chickadee checkoff" is to aid wildlife that is not helped (or may even be threatened) in the normal course of fisheries and wildlife management. All states should have such a provision—citizen action can ensure that they eventually will.

Most important, citizens, as parents and teachers, can engender in their offspring the values that lead to a concern for the future of all life and the preservation of life's diversity. The battle to save biological diversity is a battle that can be won. But not if we delay. As time goes on, it becomes a battle for our own lives, our own future. Now, while there is still time, we need to formulate a plan, a set of marching orders, to coordinate the good work now being done by so many people for so many different reasons. The plan must be a working composite, a drawing together of what may at first seem opposing goals:

1. We must encourage people to celebrate and preserve every visibly distinct form of life but meanwhile compile a short list of representative species for the few coveted spots on the zoo and aquarium ark.

2. We must create and enforce pristine habitat sanctuaries, untouched by humans and capable of sustaining as much as possible of the Earth's full complement of biological diversity. Yet we must also develop a new breed of human sanctuary in which both sustainable agriculture and sustainable cities can exist with minimal dependence on outside sources of energy and nutrients.

3. We must pull the plug on the global Waring blender of species and stop indiscriminately moving creatures from one part of the world to another. On the other hand, we must facilitate the discretionary exchange of novel crop species around the world within similar climates.

4. We must educate a cadre of bold, competent ecologists but encourage many of them to apply their insights to other professions, such as education, industry, economics, and political science.

5. We must deliver the right message to the right audience, using commercial television and film to convey the bare bones of environmental issues while delivering more informative and sophisticated treatments through print and public television and radio, so that people can vote and act intelligently.

The key to the preservation of biological diversity everywhere on Earth—in a rain forest, a coral reef, an estuary, a prairie, a city—is that people must stop thinking of all other life as the green blur out the window of a speeding train. When we stop and look at the oak trees, some important lessons are learned. There will be hope for other species when we come to regard these communities and all their inhabitants as part of our lives, along with the athletes and soap opera stars. In the modern world, no life form is too unimportant to escape our notice. The conservationists will have done their job well when no life form is too unimportant to escape our concern.

Notes

1. See the article "Scientists talk of the need for conservation and an ethic of biotic diversity to slow species extinction," *Science* 184 (1974), 646–647.

2. Norman Myers, *The Sinking Art: A New Look at the Problem of Disappearing Species* (Oxford and New York: Pergamon Press, 1979).

3. See Paul Ehrlich and Anne Ehrlich, *Extinction: The Causes and Consequences of the Disappearance of Species* (New York: Random House, 1981).

4. Catherine Caufield, *In the Rainforest* (New York: Knopf, 1985), 202.

5. Edward O. Wilson, *Biophilia* (Cambridge, Massachusetts: Harvard University Press, 1984).

6. Adrian Forsyth and Ken Miyata, *Tropical Nature* (New York: Charles Scribner's Sons, 1984). M. Balandrin et al., "Natural plant chemicals: Sources of industrial and medicinal materials," *Science* 228 (1985), 1154–1160. Ghillean Prance also discusses the potential value to human society of rain forest plants in his chapter.

7. Asa Gray, *Manual of Botany*, 8th ed. (New York: Van Nostrand Reinhold, 1950).

8. Verne Grant, *Plant Speciation*, 2nd ed. (New York: Columbia University Press, 1981).

9. Ernst Mayr, *Populations, Species, and Evolution* (Cambridge, Massachusetts: Harvard University Press, 1970).

10. Bernard Stonehouse, *The Biology of Penguins* (Baltimore, Maryland: University Park Press, 1975).

11. George G. Simpson, *Penguins: Past and Present, Here and There* (New Haven, Connecticut: Yale University Press, 1976).

12. Simpson, *Penguins*. Peter Harrison, *Seabirds* (Boston, Massachusetts: Houghton Mifflin, 1983). In fact, three subspecies of rockhopper penguins are presently recognized: *Eudyptes chrysocome chrysocome, E. c. moseleyi,* and *E. c. filholi*. Neither of our two forms conform well to the descriptions of any of these three. The only safe generalization is that there are at least three distinct rockhopper populations, and probably more than that, spread around the Southern Hemisphere.

13. George A. Petrides, *A Field Guide to Trees and Shrubs* (Boston, Massachusetts: Houghton Mifflin, 1972).

14. Carl H. Ernst and R. W. Barbour, *Turtles of the United States* (Lexington, Kentucky: University of Kentucky Press, 1972).

15. Mike Bell, "An unusual population of three-spined sticklebacks, *Gasterosteus aculeatus,* from Boston," *Copeia* 1 (1984), 258.

16. Our information base on extinction rates is incomplete because many species may be going extinct without our ever having known that they existed and also because there is little data on the distribution and abundance of many rare species. It is reasonable, however, to use species with which we are more familiar as a conservative indicator of general threats to all species, whether known to us or not.

17. Myers, *The Sinking Ark;* Ehrlich and Ehrlich, *Extinction;* David Ehrenfeld, *Biological Conservation* (New York: Holt, Rinehart, and Winston, 1970); Vinzenz Ziswiler, *Extinct and Vanishing Animals: A Biology of Extinction and Survival,* F. Bunnell and P. Bunnell, trans. (New York: Springer Verlag, 1967).

18. As a general rule, species richness goes up and abundance of any one species down as one approaches the equator. This is not true in all habitats: Much of the latitudinal diversity gradient owes to the stupendous species richness of two kinds of communities: rain forests and coral reefs. In his chapter, Prance compares the diversity of species found on small rain forest and temperate plots. For good comparative discussions of diversity in temperature and tropical forests, see Caufield, *In the Rainforest,* and Roger D. Stone, *Dreams of Amazonia* (New York: Viking Penguin, 1985).

19. Rodger A. Sedjo and M. Clawson, "How serious is tropical deforestation?" *Journal of Forestry* 81(12) (1983), 792–794.

20. Food and Agriculture Organization, Forestry Paper 30, Rome.

21. Rodger A. Sedjo, personal communication.

22. John Tierney, "Lonesome George of the Galápagos," *Science 85* 6(5) (June 1985), 50–61.

23. Robert MacArthur and E. O. Wilson, *The Theory of Island Biogeography* (Princeton, New Jersey: Princeton University Press, 1967).

24. Thomas E. Lovejoy, R. O. Bierregaard, J. M. Rankin, and H. O. R. Schubart, "Ecological dynamics of forest fragments," in *Tropical Rain Forest Ecology and Management,* S. L. Sutton et al., eds. (Oxford: Blackwell Scientific Publications, 1983); T. E. Lovejoy and David C. Oren, "Minimum critical size of ecosystems" (Lansing, Michigan: American Institute of Biological Sciences, Michigan State University, 1977).

25. G. Fryer and T. D. Iles, *The Cichlid Fishes of the Great Lakes of Africa. Their Biology and Evolution* (reproduction of 1972 ed.; Forestburgh, New York: Lubrecht and Cramer, 1982). The evolutionary radiations of cichlid fishes in Lakes Tanganyika, Malawi, and Victoria are the most expansive and rapid of any living vertebrates. Among the two hundred or more species unique to a single lake, one can find as much variation in form and function as might ordinarily be expected across ten or twenty entire families of fishes.

26. C. D. N. Barel et al., "Destruction of fisheries in Africa's lakes," *Nature* 315 (1985), 19–20.

27. Niles Eldredge and Steven J. Gould, "Punctuated equilibria: An alternative to phyletic gradualism," in *Models in Paleobiology,* T. J. M. Schopf, ed. (San Francisco: Freeman, Cooper, 1972).

28. Gould presented this speech before the 1985 Northeast Regional meeting of the American Association of Zoological Parks and Aquaria, hosted by the New England Aquarium.

29. Stone, *Dreams of Amazonia.*

30. Douglas G. Sprugel, "Dynamic structure of wave-regenerated *Abies balsamea* forests in the northeastern United States," *Journal of Ecology* 64 (1976), 889–911; Douglas G. Sprugel and F. H. Bormann, "Natural disturbance and the steady state in high-altitude balsam fir forests," *Science* 211 (1984), 390–393; Douglas G. Sprugel, "Density, biomass, productivity, and nutrient-cycling changes during stand development in wave-regenerated balsam fir forests," *Ecological Monographs* 54(2) (1984), 165–186; Douglas G. Sprugel, "Changes in biomass components through stand development in wave-generated balsam fir forests," *Canadian Journal of Forest Research* 15(1) (1985), 269–278.

31. Phillip S. Lobel, "Herbivory by damselfishes and their role in coral reef communities," *Bulletin of Marine Science* 30 (1980), 273–289. S. H. Brawley and W. H. Adey, "Territorial behavior of three spot damselfish (*Eupomacentrus planifrons*) increases reef algal biomass and productivity," *Environmental Biology of Fishes* 2 (1977), 45–51.

32. Leslie S. Kaufman, "The three spot damselfish: Effects on benthic biota of Caribbean Coral Reefs," in *Proceedings of the Third International Coral Reef Symposium* (Miami, Florida: Rosenstiel School of Marine and Atmospheric Science, 1977); Leslie S. Kaufman, "Damselfish disturbance on Caribbean coral reefs," Ph.D. dissertation, Johns Hopkins University, 1979.

33. Leslie S. Kaufman, "There was a biological disturbance on Pleistocene coral reefs," *Paleobiology* 7 (1981), 527–532.

34. Robert Endean, "Population explosions of *Acanthaster planci* and associated destruction of hermatypic corals in the Indo-West Pacific Region," in *Biology and Geology of Coral Reefs*, O. A. Jones and R. Endean, eds. (New York: Academic Press, 1973), vol. 2, section 1.

35. Ann M. Cameron and R. Endean, "Renewed population outbreaks of a rare and specialized carnivore (the starfish *Acanthaster planci*) in a complex high diversity system," in *Proceedings of the Fourth International Coral Reef Symposium* (Manila: Paragon Printing Corp, 1981), vol. 2. H. Mergner, "Man-made influences on and natural changes in the settlement of the Aqaba reefs," in *Proceedings of the Fourth International Coral Reef Symposium* (Manila: Paragon Printing Corp, 1981), vol. 2.

36. Edgar Frankel, "Previous *Acanthaster* aggregations in the Great Barrier Reef," in *Proceedings of the Third International Coral Reef Symposium* (Miami, Florida: Rosenstiel School of Marine and Atmospheric Science, 1977).

37. Randy Olsen, personal communication.

38. Mitchell W. Colgan, "Succession and recovery of a coral reef after predation by *Acanthaster planci* (L)," in *Proceedings of the Fourth International Coral Reef Symposium* (Manila: Paragon Printing Corporation, 1981), vol. 2.

39. J. D. Woodley et al., "Hurricane Allen's impact on Jamaican coral reefs," *Science* 214 (1981), 749–755.

40. H. A. Lessios, D. R. Robertson, and J. D. Cubit, "Spread of *Diadema* mass mortality through the Caribbean," *Science* 226 (1984), 335–337.

41. E. B. Stephens et al., "A viral aetiology for the annual spring epizootics of Atlantic menhaden *Brevoortia tyrannus* (Latrobe) in Chesapeake Bay," *Journal of Fish Diseases* 3 (1980), 387–398.

42. Martin Newman, personal communication.

43. Stephens et al., "A viral aetiology."

44. Caufield, *In the Rainforest*, 81.

45. Rupert Russell, *Daintree: Where the Rainforest Meets the Reef* (Australia: Kevin Weldon and Associates and the Australian Conservation Foundation, 1985), 179.

46. Caufield, *In the Rainforest*, 74.

47. Russell, *Daintree*.

48. Barel et al., "Destruction of fisheries in Africa's lakes."

Appendix

The Last Extinction was published seven years ago. Since then the world has changed beyond belief. The world economy boomed and then felt the sickening tug of gravity. Communism has been relegated to the long sleep of the library shelves. Nuclear conflagration, though still possible, is less probable. Into the political and economic vacuum have come thrilling new things. Biotechnology is an industry now, not just a risk. A fleet of new telescopes in the sky and on the ground have revealed haunting images of fetal stars, new planets, and black holes.

Along with these wonders has come the broad sharing of some very important human feelings. We are alone, save for each other. Life is precious and rare, and this planet has all there is as far as we're concerned. We had better take care of it. The first world summit focusing on the environment, the United Nations Conference on Environment and Development (UNCED), was held in Rio De Janeiro in the summer of 1992. The planet is abuzz with stories, books, films, television, toys, meetings, and good acts about and toward nature. This is a new world. This is the new chance.

The world has changed in astonishing ways, but also in ways that fulfilled our worst expectations. Efforts to curtail population growth have failed on the whole. The world market is swelling, but most cannot buy, and many cannot eat. The limits to growth have been pressed until pressurized. Death stubbornly shadows vast regions of Asia and Africa. AIDS is joined by an army of new pathogens, including resistant, educated strains of flu, malaria, and bilharzia.

In the meantime, the human-damage-unit (HDU)—that is, the potential for one human to wound the environment during his or her lifetime, has hit an all-time high. Recall the incinerated, day-for-night oil fields of Kuwait; poisoned pathways through the former communist world; bare earth and fuming fires in the Amazon. Or just look up through the unprotecting atmosphere—but wear dark glasses and a hat. Maybe things didn't really change so fast, maybe we're just suddenly more aware of them. But the situation has definitely been altered.

Picking up where this chapter left off seven years ago, one can cite a host of encouraging changes. The informed public is now well aware of the plight of endangered species and habitats. Much movement is afoot to save the rain forests. The Australian rain forests have achieved World Heritage designation, being thus declared areas of outstanding universal values and protected by decree of the International World Heritage Committee. The Plymouth red-bellied turtle, the Willow Pond stickleback, the rockhopper penguins, the striped bass, the wild Atlantic salmon—they're all still eking out an existence, and striped bass have even had a couple of banner years.

During these seven years, the New England Aquarium has led a group of other aquariums around the world in organizing breeding programs for endangered fishes, joining other private, state, and federal efforts. The California condor, a bird that some folks said would never again paint the sky, is once more poised upon thermals in the wild, albeit tentatively. Those species lucky enough to have received such coddling have sometimes done well. It is all very heartening—but ultimately misleading. For while we've undoubtedly helped a few species to cheat death, the very fabric of their being, the living landscape that spawned and nurtured them, is changing or fading away.

In the sea, the signs of dying ecosystems are clear. The major fisheries of the world are all in trouble, a matter that came home to roost when the Georges Bank fishery shrank and was forced to change the targets of the hunt from commercially valuable cods and flatfishes to less desirable trash species such as dogfish and skates. The giant bluefin tuna has declined more than 90 percent in twenty years, a drop comparable to that of the African elephant. Populations of the world's whales remain threatened, as you will discover in Norman Myers's new chapter for this volume called "Sharing the Earth with Whales."

The situation is even worse in freshwater environments, subject to the concentrated impacts of entire watersheds (see update by Williams and Nowak). In aquariums and zoos, endangered species in need of temporary arks are replacing other more familiar animals in the collection. At the New England Aquarium we must send new infusions of captive-reared Plymouth red-bellied turtles regularly back into the wild, and together with twenty-nine sister institutions we maintain and breed over forty species and subspecies of Lake Victoria cichlids. Meanwhile, however, all the rest of the native aquatic fauna of Massachusetts—all the supposedly nonendangered species—are in deep trouble too.

The alarming fact is that about one-fourth to one-third of all freshwater fishes around the world may now be threatened or endangered, a story of devastation that is most poignantly told by the account of Africa's Lake Victoria. Within a single decade more than two hundred species of its fishes have disappeared from the lake—half of the endemic fish fauna, now presumed extinct (Witte et al. 1992).

The destruction of Lake Victoria's fish community isn't the greatest extinction in which humans have been indicted. The wipeouts of large land animals in the Pleistocene were more dramatic and covered more ground, if fewer species. But bones and a few cave paintings are all that we have for a record of that event. The list of insects and plants forever gone from the world's tropical forests would probably number hundreds of thousands of species—if they hadn't disappeared before we even knew that they were there. Lake Victoria is special because it is the site of the first mass extinction ever witnessed, and felt, and intensely analyzed.

Details of this gory saga are to be had elsewhere (Kaufman 1992, Witte et al. 1992), but two points are worth mentioning here. The first is that these extinctions are the result of a combination of factors, not just one. Overfishing, alien introduction, nutrient pollution (possibly including acid rain), and defor- estation all contribute to the collapse. Second, the extinctions in Lake Victoria are interconnected, just as the species were in life. Varied factors are conspiring to undermine the integrity of the lake ecosystem as a whole.

The lake is now plagued by algal blooms, oxygen depletion over 50 percent to 70 percent of the water column, deadly upwellings and fish kills, and intense predation by Nile perch, an introduced predator, on anything that moves in shallow waters. Significantly, among the hundreds of species lost are several of enormous value as food for people living near the lake. The Nile perch is both undesirable and too expensive for local tastes, supporting instead a major export market. The result: local protein deficit in the midst of un- precedented fish production. So, of the vast evolutionary carnage piling up over the past seven years, the mess in Lake Victoria is most significant as the one we can best see and feel. It is the Hiroshima of the biological apocalypse, the demonstration, the warning that more is on the way.

Another small part of our ecological coming of age is that nothing on the planet is remote any longer. When it appeared that only penguins would suffer from the effects of chlorofluorocarbons (CFCs) on the ozone layer, it didn't seem to pose the same imminent threat to the rest of us. This attitude changed, abruptly, with the opening of an ozone-depleted region directly over Europe and North America. Now the use of CFCs is finally being reigned in, and rates of ozone erosion have graciously followed suit.

The relationship between human activity and global warming is more controversial. Be this as it may, it is not theory that the west Atlantic continues to be plagued by a succession of extraordinary hurricanes. Hurricane Allen was followed by Hugo, Gilbert, and Andrew, the latter slashing southern Florida with a narrow but ferocious track likened to that of a 20-mile-wide tornado.

The effects of such hurricanes vary greatly with habitat. The famous Florida Everglades took a direct hit. But most Everglades communities are adapted to, and may even have benefited from the onslaught. One precious habitat that did not benefit is the tropical hardwood hammock, the major community type on dozens of tree islands that once dotted the eastern glades. The best remaining hammocks, prior to Andrew, were in West Homestead and Miami, surrounded by suburbs and farmland.

As discussed earlier for Jamaican coral reefs, vulnerability to natural dis- turbance can be greatly increased by prior human impacts. Years ago, the hammocks were destabilized by alien vines that erupt each time a tree falls, preventing subsequent regeneration by native tree species. Hurricanes like Andrew merely accelerate the system-wide transformation into rolling blankets of weed and vine, with here and there old trees poking through, hanging on a bit longer.

In a different way, the storm was not bad enough to do good. Andrew missed Florida Bay, thus failing to flush out the stink of the eutrophication that had accumulated over decades. Deadly nutrients are the legacy of overdevelopment and underplanning, as well as a big contributing factor to the continued decline of Florida's "living coral reef." Soon, the tourist industry that depends upon them must decline as well.

What we have witnessed in the past seven years is the Ehrlichs' rivet popper analogy made real. When enough species are removed from a system, it really does undergo change: convulsive, radical change. With so many species gone, systems collapse and simplify, stabilized in new conformations that we will be hard put to alter. In some places, like Lake Victoria, this is happening so quickly that even those species that survive that first wave of change are left without home or context. Time itself has ended for them.

Noble efforts have been mounted to combat the loss of species once supported by these failing ecosystems. The struggle is underway to save sturgeons in the creeks that still feed the parched plain that was once the Aral Sea. Scientists are rescuing rainbowfishes and lemurs on a desolate Madagascar, bluefin tuna in a ravaged North Atlantic, and lizardlike tuatara clutching some God-forsaken rocks off New Zealand. Beneath all this dedication and sweat and blood there is this inescapable image: so many building janitors rushing about catching light bulbs falling from a huge ceiling in the midst of an earthquake. They will be left with a collection of light bulbs, all right, but with no sockets to screw them back in to, maybe no building at all.

Everywhere, ecosystems are changing so fast that the remaining orphaned species search for new homes that are too few, too small, and too far away. The times are a tragicomic counterpoint to the old, heated arguments between species versus habitat preservationists. It is altogether too late for preservation alone to ensure a livable world. It is even too late in most places to try to restore what was there before: huge chunks of the flora and fauna have been lost forever. What is needed now is a parallel track to preservation—a totally different perspective on species conservation.

Historically, species conservation programs have been considered worthless unless they fulfilled the goal of restoring species to their original habitats in the wild. Some felt that when the point was passed at which this could occur, the species should be allowed to go extinct without prolonging the agony of the few remaining individuals. Seven years ago, I almost believed this nonsense.

If we adopted this strategy now, there'd be no point in saving anything. The entire context of life has been rewritten. This is not your parents' planet.

So what, then, is our heritage? It is a world of shattered ecosystems. It is aquarium tanks, animal cages, and botanical hothouses full of homeless species. It is government files full of paper parks never realized. In a lot of places it is already much too late for conventional strategies to be of use. If systems are so severely disturbed that their chief occupants are ecological waifs, then the only hope for saving these species in the wild is to disturb the system yet again and reinsert large numbers of species in the hope that they will create a new set of functional interrelationships.

Not that we know how to do this yet. It's like one of those awful toys that are all pieces, no instructions. Assembly can be very disheartening unless you know the secret. The secret is, it doesn't matter what you come out with as long as it integrates most of the pieces and allows a harmonious place for people in the system. If you can only succeed in this, at least your options have been preserved. If you can't restore, you reconstruct. Sometimes, you wind up with something quite different than what you see in the old pictures. Just remember, it arrived at its original state in much the same way: a big dose of happy accident.

Ecological reconstruction does have a few rules. First, the original pieces must be valued above all others. They fit together once before, and are much more likely to fit together again than stray elements from a different kit. Second, you need the framework or habitat: trees, coral, organisms that build habitats, produce food, and moderate the physical environment are of crucial importance. Third, you must try the experiment over and over again, because often it won't work. When it does, it will rarely work twice in precisely the same way. Fourth, you must give it time. Lots of time. Eventually you will learn how to set things up so they are likely to assemble into a system that can sustain itself. The final rule is this: the goal is to create something that has never existed before. The goal is create a living community in which we can live harmoniously with all other species. Otherwise the exercise is pointless.

Where we are luckiest, we will be able to reconstruct scenarios not too unlike the ones we laid waste, and sometimes even better than they've been in recent memory. Elm and chestnut can someday be resurrected to a new eastern deciduous forest in North America. Lake Victoria can have back most of the food fishes lost during the first half of the century—they're sitting, waiting in hatcheries and ponds and laboratories right now. Georges Bank can again produce abundant cod and haddock. Florida can again be renowned for its crystal springs, lush hardwood forests, and glowing coral reefs. East Africa can be a paradise for game once more, and the Great Plains can be returned to some semblance of greatness. And all this can happen within reason, with due regard to the seething mass of humanity pushing at the park gates and the fishing grounds. That's the one advantage of things getting so bad—soon nobody will remember how good they were.

The call to arms? Revitalize the world! Bring back the vigor of natural communities, and find nestled in their midst more ways to make good lives for fewer people. And in anticipation of this day, which I fear is still a good distance away, preserve what can be preserved, and nurture it. Somewhere between the results urged by the ardent preservationist and the bold reconstructionist, we shall forge a new home for ourselves on this earth. Or die trying.

★   ★   ★

This update is dedicated to Dr. John Sutherland of Duke University, whose work was influential in effecting change in ecologists' thinking from the idea of linear succession to the notion of living communities shifting among "alternate stable points."

References

Y. Baskin, "Africa's troubled waters: Fish introductions and a changing physical profile muddy Lake Victoria's future," BioScience 42 No. 7 (1992), 476–481.

L. S. Kaufman, "Catastrophic Change in Species-Rich Freshwater Ecosystems: The lessons of Lake Victoria," BioScience 42 No. 11 (1992), 846–858.

F. Witte et al, "The destruction of an endemic species flock: quantitative data on the decline of the haplochromine species from the Mwanza Gulf of Lake Victoria," Environ. Biol. Fishes 34 (1992), 1–28.

2

Mass Extinctions: New Answers, New Questions

David Jablonski

Paleontologists have known for over a century that changes in the 4-billion-year history of life have not always been slow or steady. Major revolutions, such as the disappearance of the dinosaurs or of the ammonites, can be clearly seen in the fossil record. These events are termed mass extinctions. The geological and paleontological discoveries of the past five years have led to a new look at the phenomenon of mass extinction, both causes and consequences, and indeed a new look at the history of life on this planet. We certainly do not have the full story, but recent discoveries have set us on the track of a whole new set of problems and questions. This chapter is a progress report, not a final summary. It contains a brief overview of research on mass extinction, speculates about where the field is going, and considers the lessons of the geologic past for life on Earth today.

Mass extinctions have plagued the Earth's biota repeatedly over at least the past 600 million years, the time interval since the origin of well-skeletonized organisms that make up the richest part of the fossil record. Geological, geochemical, and paleontological evidence suggests that one or more of these mass extinctions was either triggered or intensified by impacts of extraterrestrial objects. The role of these mass extinctions in shaping the history of life is still poorly understood, but new evidence suggests that they did not result simply from intensification of natural selection and other evolutionary processes that prevailed during the long time between mass extinctions. Instead, the rules of extinction and survival apparently change for a geologically brief period of time, removing many groups that are well adapted to the normal or "background" extinction processes that have prevailed over most of the Earth's history. The victims of mass extinction thus might easily include groups that were dominant during normal times, whereas survivors might include groups normally vulnerable to extinction.

The history of life may thus be an alternation of normal and mass extinction rules, with a given group having radically different levels of vulnerability at different times in Earth history. This is a far cry from the more comfortable view that extinction serves to weed out the weak or the poorly adapted, making way for better-designed, more-advanced organisms. Particular groups of animals might rise to dominance, not because of any relative superiority but because they happen to weather a brief and unpredictable crisis. The great mammalian radiation, including our own lineage, may be the best example of all. Mammals may well owe their present dominance to a short-lived shift in the survival rules rather than to any innate superiority that mammals had over dinosaurs.

What Is a Mass Extinction?

Extinction is one of the harsh realities of the history of life. Unfavorable changes in local environments, bad luck over a series of crucial breeding seasons, and many other factors maintain a fairly steady background level of "natural" extinction, averaging perhaps a few species per million years for most kinds of organisms. Authors use different criteria to distinguish time intervals that require special attention. One of the best definitions of a mass extinction is one that doubles the extinction background level among many different kinds of plant and animal groups.[1,2] Thus a mass extinction involves a simultaneously heightened extinction among, for example, bottom-dwelling clams and surface-floating plankton or among land-dwelling dinosaurs and ocean-going mosasaurs. The accelerated extinction among many groups at once is important for this definition: when such independently evolving systems lock step, we have a sign that something extraordinary has happened.

Using these criteria, we can see at least five mass extinctions in the fossil record, and more detailed study could ultimately reveal a dozen or more. Figure 2.1 shows John Sepkoski's plot (from his 1984 article; see note 2) of the number of families of readily fossilized marine animals—the groups for which we have by far the most reliable record—since the diversification of complex, skeletonized forms at the beginning of the Paleozoic Era about 600 million years ago. These five biggest extinction events show up clearly as sudden drops in the number of families present in the oceans, followed by somewhat less rapid recoveries that last millions of years each. The late Permian extinction is the most devastating by far, having removed over 52 percent of the marine families. Families are large evolutionary units, and estimates that trace the size of the observed familial extinction to the species level suggest that a staggering 91–96 percent of the then-living species were lost!

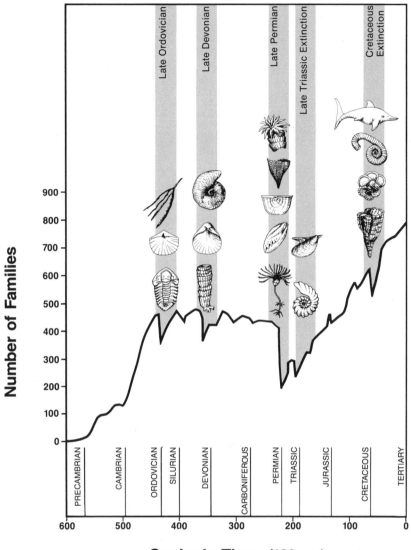

Geologic Time (10⁶ yrs)

Figure 2.1

A summary of mass extinctions showing the principal groups of animals affected by each event. These groups are (from top to bottom) graptolites, articulate brachiopods, and trilobites (Late Ordovician); goniatites, brachiopods, and corals (Late Devonian); corals, bryozoans, brachiopods, pennoid foraminifers, and crinoids (Late Permian); bivalve mollusks and heteromorphic ammonoids (Late Triassic); and ichthyosaurs, heteromorphic ammonites, planktonic foraminifers, and rudist bivalve mollusks (Cretaceous). Original illustration by Carol Bayle, based on figure 2 in D. M. Raup and J. J. Sepkoski, Jr., "Mass extinctions in the marine fossil record" (*Science* 215 (1982), 1501–1503). Reprinted by permission.

Although not quite as massive, each of the other four major extinctions represent the disappearance of about 20 percent of the families present and presumably a much larger proportion of the species. All these extinctions— including the devastating one at the end of the Permian—are so distant in the geological past, however, that we have no way of knowing how rapidly these events took place. The extinctions almost certainly took less than five million years, probably less than one million years, but resolution begins to fail beyond that level of precision. Some scenarios for impact-caused extinction allow only a few years, or even months, for the entire process to take place; other hypotheses involve tens to hundreds of thousands of years for a mass extinction to occur. Direct geological evidence that would allow us to decide on the time scale of these events is not yet available.

Our imprecise understanding of how long ancient mass extinctions lasted makes it difficult to draw direct comparisons between the state of today's beleaguered biosphere and the upheavals of the geologic past. Nevertheless, authors such as Paul and Anne Ehrlich and Norman Myers believe that by the year 2000 human-caused extinction will have accelerated to Late Cretaceous or even Late Permian levels, adding a new extinction trough to the jagged diversity curve in figure 2.1. As I discuss later, over the past few centuries and in the century to come, just a geological instant, our species will probably cause the worst disaster in the past 65 million years for life on this planet.

The Late Cretaceous Extinction

Although the Late Permian extinction was the most severe, the one at the end of the Cretaceous has enjoyed the most publicity. Not only are the victims of the extinction—the dinosaurs and other spectacular reptiles—a source of perpetual interest, but the late Cretaceous record has recently yielded a wealth of evidence that asteroid or comet impacts could have triggered the mass extinctions. Almost everyone has emphasized the victims of the event, but a more balanced approach that takes stock of the survivors as well can help to test ideas about causes of the extinction and decipher the evolutionary significance of the event.

The most celebrated animals to vanish at the end of the Cretaceous were, of course, the dinosaurs. Apparently all evolutionary lineages of dinosaurs, about fifteen families, became extinct about 65 million years ago—herbivores, carnivores, scavengers, huge animals far bigger than elephants, and tiny animals no larger than turkeys. Both families of pterosaurs, flying reptiles related to dinosaurs, disappeared at the same time. A remarkable and varied assemblage of animals was lost, but many other aquatic and land-dwelling animals were left almost unaffected, including large reptiles, such as the crocodilians—still with us today—and the crocodilelike champsosaurs, which persisted for another 25 or 30 million years. Our distant ancestors, the small placental mammals, were left virtually unscathed, as were snakes, lizards, turtles, and freshwater clams and snails. The fossil record of birds is too sparse to evaluate for ancient extinctions, but it is clear that this particular branch of the dinosaurian lineage did survive.

The record of terrestrial plants also presents a complicated picture; it was not wholesale massacre. Although the fossil record of leaves, twigs, and flowers is not complete for this time interval, pollen grains are extremely abundant, and they reveal an interesting geographical bias in extinction. A fossil assemblage called the *Aquilapollenites* province, named for the dominant pollen type, shows a major changeover across its entire range of western North America, Siberia, and the Far East. But the *Normapolles* province in eastern North America and Europe shows few signs of disruption. Similarly uneven levels of extinction appear in the less-well-known pollen provinces of the Southern Hemisphere.[3] Overall, when compared with the record for terrestrial animals, the plant fossil record shows remarkably little extinction above the species level: shifts in geographic ranges, presumably in response to changes in climate, appear to be the norm. Some feel that this lack of a profound mass extinction argues strongly against a truly catastrophic event at the end of the Cretaceous, but Andrew Knoll[4] has suggested that even then plants could show a high survival rate because of their great regenerative powers. Thanks to their roots, tubers, and seeds, plants might experience only minimal extinction even in the face of total above-ground defoliation.

The best evidence for abrupt extinction is in the ocean, among the unicellular plankton, where the two most readily fossilized groups underwent an impressive crash. The coccolithophorids, unicellular algae whose calcium carbonate shells produced the White Cliffs of Dover and many other chalk deposits that give the Cretaceous Period its name, show a sudden and major extinction at the species and genus level. And the planktonic foraminifers, major protozoan contributors to calcareous deepsea sediments, almost did not make it: only one or two species appear to have survived the mass extinction, although they rediversified into a wide array of new species soon after. Unfortunately, other important plankton groups, including dinoflagellates, radiolarians, and diatoms, lack a sufficiently detailed fossil record to provide much more information on this extinction.

Larger marine animals, such as the reptilian plesiosaurs, ichthyosaurs, and mosasaurs, also became extinct during the Cretaceous, as did the ammonites, a fantastically diverse lineage of mollusks related to the chambered nautilus, and the belemnites, relatives of the squids and cuttlefishes. But it is worth noting that some of these groups were not especially diverse at the time of their final demise. The ichthyosaurs seem to have been reduced to only a few species, and even the ammonites, which certainly lost a large number of families, had already dwindled considerably compared with their heyday during the mid-Cretaceous, some 35 million years before their extinction. The same can be said for the belemnites.

Bottom-dwelling, shelled marine invertebrates have a diverse and well-preserved fossil record, and their evolutionary patterns are often used as a barometer for extinction and the development of new life in the oceans. The most conspicuous victims here were the rudists, massive tropical clams that had replaced the corals as the prime reef builders in Cretaceous seas. Lost along with this group were the nerineacean snails, large tropical mollusks with elaborate shells. The sea urchins and the sponges also lost considerably more families than expected from background levels of extinction. Outside the tropics, however, losses of clams, snails, and bottom-dwelling forminifers were relatively minor.

We are left with a complex portrait of the late Cretaceous extinction—not simply the elimination of the dinosaurs and other huge reptiles but not utter annihilation either. Tropical marine groups were more severely affected than temperate or polar ones; open-ocean plankton and larger swimmers, such as the mosasaurs, were affected more than bottom dwellers; and large land dwellers more affected than small ones, even though some larger forms survived as well. How does this complicated picture fit into current ideas about the causes of the extinction?

There is a litany of unsubstantiated and sometimes absurd explanations that abounds in the scientific and popular literature, from dinosaurs with terminal constipation because they ate newly evolved flowering plants to the handiwork of egg-eating mammals. Both such mishaps might well have done the dinosaurs in, but they do not explain the simultaneous extinction of, for example, the planktonic foraminifers or the *Aquilapollenites* flora.

One mechanism for extinction that we could have dismissed as yet another farfetched hypothesis seven years ago has since risen to prominence: the Alvarez hypothesis that the impact of an extraterrestrial object triggered the Late Cretaceous extinction. Whether this idea is correct or not, it has been immensely valuable to the study of mass extinctions because it brings a whole class of sudden-event, catastrophic causes into the realm of testability. If a hypothesis can be framed so that it can be tested, that is, if it specifies some particular observation or test that can prove it to be wrong, then it can be treated scientifically. And the impact hypothesis does just that.

The Alvarez hypothesis took shape with the discovery of anomalous amounts of iridium in a clay layer exactly at the level of the Late Cretaceous extinction. Iridium and other platinum group elements are scarce in the Earth's crust: the Earth's primordial supply presumably sank into its interior 4 billion years ago as our planet cooled and differentiated from its homogeneous molten state. Thus a hundredfold enrichment of iridium could be taken as evidence that some kind of extraterrestrial impact, an asteroid or comet whose core was similar to the solar system's original composition, occurred at the end of the Cretaceous and was responsible for the extinction.

The impact idea has generated heated discussion, but more important, it immediately suggested a number of potential tests: explicit predictions that, if not met by new observations, would undermine the hypothesis. One such prediction is that the iridium-rich clay layer should be present, exactly at the level of the extinction, wherever there is a complete sedimentary record. Continuing research has borne out this claim. Virtually every locality that has been checked, on land and sea, from Denmark to Tunisia, Montana to the central North Pacific and New Zealand, has the expected anomaly in just the right place.[5]

Another prediction is that the crucial layer should contain other by-products of an impact. Geochemically this has been found to be the case: Osmium, chromium, cobalt, and other elements are more abundant in that layer than should be expected in the Earth's crust. There are also physical clues: tiny spherules a few tenths of a millimeter in diameter that closely resemble microtektites, which are known to form during impacts, and sand grains in which the quartz has been stressed—as if by a great impact—into unusual minerals and crystallographic forms.

By meeting these predictions, the impact hypothesis has withstood the first round of testing and deserves to be taken seriously. Not everyone accepts it, however. Perhaps the most important challenge to the idea of Late Cretaceous extraterrestrial impact comes not from doubting whether the evidence is there, but from questioning whether the evidence could have been produced only by an impact. Certain kinds of volcanoes release iridium and other unusual elements and may even produce stressed quartz, lending some support to an alternative hypothesis involving major, deep-rooted volcanic eruptions of a type known to have occurred near the end of the Cretaceous.[6] As of this writing, the bulk of the evidence still seems to weigh toward some kind of extraterrestrial impact, but the issue remains undecided, and subtle geochemical and mineralogical arguments on isotopes and dust-deterioration products in extraterrestrial versus volcanic debris have taken on a new urgency.

Even if we accept that an impact did take place, the question remains whether it caused the mass extinction, and if so, how. There are plenty of mathematical models for the effects of an impact that suggest it would do the job—and quickly. But there is something disquieting about the spectacular array of disastrous effects that the modelers have generated.[7] The anticipated effects are so devastating that they seem inconsistent with the complex biological patterns just outlined. Various authors have argued that impact by an extraterrestrial object sufficiently large to leave so much iridium evenly spread over the Earth's surface might:

• Produce so much dust that there would not be enough light to support photosynthesis for two months to a year. With the cessation of growth among land plants and planktonic marine algae, food chains would collapse, and most animals would starve.
• Produce so much dust that sunlight could not warm the Earth's surface. Temperatures in the inland regions, including the tropics, would drop below freezing for six months to two years. This would trigger a mass mortality of animals adapted to mild climates.
• Create, because of the shock of impact, a host of dangerous chemical reactions in the atmosphere, yielding a deadly global rain of hot nitric acid.
• Throw so much water vapor into the atmosphere, in the likely event that the impact occurred somewhere on the 75 percent of the Earth's surface that was ocean at the time, that it would create a greenhouse effect with lethally high temperatures.
• Produce an immense tsunami, a wave at least 300–400 meters (1000–1200 feet) high, which would drown all low-lying land areas, inundating all but the mountain ranges and silting over and killing most vegetation. Those dinosaurs that did not drown would starve.

This partial list is an impressive doomsday menu, but it does not seem consistent with the complicated pattern of the extinction. Scientists have tended to emphasize the victims and have been neglectful of the survivors. Such emphasis on the victims alone fails to explain why a low proportion of freshwater animals became extinct or why many nontropical marine invertebrate families survived. And it does not explain the lack of any geological evidence for a massive silting-over of the continents. These suggested catastrophes require the extinction to be an extremely brief event, lasting a maximum of ten years.[8] On a geological time scale events that last one year or one thousand years are almost inseparable geological instants.

There is evidence from some of the most complete fossil sequences suggesting that the extinction, and possibly even the deposition of iridium, may not have been geologically instantaneous (see Officer and Drake's article in note 6). And I cannot shake my own (unsubstantiated) feeling that a disaster powerful enough to kill every dinosaur, ammonite, and reef-building clam in a single decade, or a single year—or even one bad weekend!—would not allow so many other animals and plants to slip through unscathed.

This is not to dismiss an impact as the culprit at the end of the Cretaceous Period. We have so little to go on when we try to estimate what such an event would really do to the Earth that it would be a mistake to assume that the impact effects proposed so far are definitive. A broad spectrum of effects is possible. For example, impacts could trigger climatic changes lasting thousands of years by disrupting stratospheric circulation patterns or by altering the cloud cover or the water-vapor content of the atmosphere. Such effects might be more subtle, but they are potentially just as far-reaching as the "lights out" or "deep freeze" scenarios. They are also more consistent with the available geological and paleontological evidence. A new generation of models for impact effects will doubtless emerge in the next few years. And new, definitive evidence supporting or disproving much of what I have written here may turn up at any time.

Comparing the Mass Extinctions

The extinction at the end of the Cretaceous is only one of many crises in the history of life, and one way of looking at mass extinctions as a general problem is to compare them. Common denominators might help us to get a better feel for the causes and the consequences of these events.

Ice ages, global climatic extremes that produce major glaciations, are often the first factors that come to mind. But explaining mass extinction by glaciations does not work. There is a close correspondence between the Late Ordovician extinction and glaciation, and the same relationship could apply to the Late Devonian extinction. A major glaciation during the Carboniferous, however, lasted almost 90 million years, and there was no mass extinction accompanying it. Furthermore, there is no evidence of major polar ice caps at the ends of the Permian, Triassic, and Cretaceous periods, let alone large-scale glaciations. And so far as we know, the onset of an ice age several million years ago brought little extinction. Thus, even if some sort of climatic change was involved during the mass extinctions,[9] as suggested by the fact that tropical marine faunas, especially reef dwellers, are always most severely affected in extinctions, ice ages were not a general cause.

All the major extinctions affected both marine plankton and bottom dwellers. We can therefore rule out sudden increases in ultraviolet or cosmic radiation produced by solar flares or nearby supernovas. Such events might be devastating to land dwellers and to marine plankton close to the sea surface, but the shielding effects of seawater would protect the deeper plankton and the bottom dwellers.

All mass extinctions were most severe in the tropics, as least for marine animals. In every instance, the reef community was virtually destroyed, and it usually took millions of years before the next reef community got underway. Following the demise of the rudist clams at the end of the Cretaceous, for example, there was a 10-million-year hiatus in reef building before the coral community was reassembled. In contrast, high-latitude marine communities usually fared much better. Overall, the pattern suggests a climatic disturbance, detrimental to tropical communities but less devastating to those in the already seasonal and cooler temperate and polar regions.

Marine regression—the retreat of the sea—is the most impressive common denominator in terms of environmental change during mass extinctions. This particular correlation between mass extinctions and drops in sea level is a constant headache for paleontologists, because regression also destroys the fossil record. It allows erosion during exposure of the continental shelf when the sea level rises again. The record of the extinction time interval is thus maddeningly difficult to decipher in detail.

The simplest explanation for the coincidence of regression and mass extinction, at least for marine animals, is the direct effect of reduced habitat area. As the habitable sea floor decreases, smaller and fewer populations can be supported there, and extinction of many species will be the inevitable result. This argument is elegant, but it does not stand up for at least two reasons. In the first place, although every extinction is associated with a marine regression (with the possible exception of the Late Devonian), not every marine regression has a mass extinction. There is no room in this simple, mechanistic explanation for so many exceptions.

Second, drops in sea level may expose great areas of continental shelf, but they will have little effect on the habitats around oceanic island chains. These chains consist of conical or drowned, flattop volcanoes, so that a drop in sea level will actually increase the perimeter of shallow-water area ringing the conical island slopes. The inhabitants of oceanic islands should thus gain, not lose, habitat area during regression. We have only a sparse fossil record from oceanic islands because it is swallowed up in deep-sea trenches when the crustal plates bearing the islands converge. So, to test the potential effect of these island refuges, I made a census of shallow marine faunas living today on oceanic islands.[1,10] It turns out that oceanic islands harbor a remarkably large proportion of today's marine invertebrate families. Even if we assume the worst case, that regression eliminated every living thing from the continental shelves, fully 87 percent of the marine families would survive on islands. The island refuge is so species rich that a reduction of continental shelf area alone could not produce the massive extinctions observed in the fossil record.

There are, however, several other possible effects of regressions that might link them to mass extinction. Regressions can destroy unique habitats on land and sea and can trigger climatic changes; the severity of both effects probably depends on the relationship between land and sea before the drop in sea level. If sea level is high, then a small vertical drop will not only expose large areas of seafloor but also destroy unique marine habitats, because shallow seas that cross the interiors of continents harbor unusual oceanographic and ecological conditions. The importance of the starting configuration might help to explain why the Late Ordovician glaciation brought on a mass extinction, whereas the most recent glaciations have not. Sea level was particularly high during the Ordovician Period, so that only small areas of land were above water on most continents, including North America. When the Late Ordovician glaciation began, vast and unique inland sea habitats were destroyed as the seas lost water to the ice caps and shorelines retreated to the edges of the continents. In contrast, the seas were already at the edges of the continents when the recent ice ages of the Pleistocene began. Only a ring of shallow-water habitat was lost when sea level dropped.

Land environments are also disrupted when sea level drops and regression occurs. Rivers carve new, deeper streambeds and break up floodplains, which would be lost over vast areas with the withdrawal of inland seas. All the eroded sediment would be dumped into nearshore marine environments, leading to yet more habitat changes. This new load of sand and mud would do most harm in the tropics, where reef builders depend on clear, sediment-free waters to allow the living reef animals to feed, grow, and receive light for photosynthesis.

Most important of all, a drop in sea level from a starting position high on the continents would have major effects on global climate, bringing a shift from a mild, Mediterranean climate, like that of central California, to a more severe, seasonal climate, like that of central Iowa. The result would be changes in annual frosts, rainfall, summer heat, day/night extremes, and many other factors of great importance to marine and terrestrial animals alike, and extinction is a plausible outcome. However, plausibility is not proof, and these hypothetical chains of cause and effect are difficult to test. Slow progress is being made as paleontologists, geologists, and climatologists grapple with different parts of the problem. If asteroid or comet impacts can produce the same climate changes as marine regressions, as I speculated earlier, then the problem of generating separate predictions to test one possible extinction trigger against the other becomes especially difficult.

The latest twist to the mass extinctions story suggests that it might be a mistake to separate completely the effects of impact and regression: perhaps both mechanisms are important, and the biggest mass extinctions occur when they combine. David Raup and John Sepkoski[11] took a new look at the diversity curve for marine families over the past 250 million years, where the time scale has the best resolution. To their astonishment, they did not find the random string of peaks and valleys that one might expect but rather a succession of periodic extinctions, with an event coming approximately every 26 to 28 million years. The most recent one was, thankfully, only 13 million years ago, so we are now safely in the middle of the cycle. Raup and Sepkoski's finding has sparked a new round of controversy and research. Many paleontologists and statisticians are skeptical; they are not convinced that all the supposed extinctions really stand out above background or that there really is a strong periodic pulse to the extinction curve. The controversy still rages among these specialists.[12] The astrophysicists, on the other hand, who agree with Raup and Sepkoski that it is hard to imagine causes intrinsic to the Earth or its inhabitants, give their enthusiastic support, along with numerous proposals for driving mechanisms.

Astronomers soon focused on the cloud of comets surrounding the Solar System. Any force that disturbs that cloud at regular intervals could send a shower of comets hurtling into the inner Solar System. A few of the comets would certainly hit the Earth, thus triggering the Late Cretaceous and all the other extinctions. There was an early suggestion that such comet showers might be set off as the Solar System wobbles in and out of the plane of the Galaxy, but this idea has since been discarded (see note 12). The two remaining proposals, of a tenth planet or of a dim, small star circling the Sun, are also embattled and may not survive mathematical testing.

One prediction of the comet impact idea, however, has been borne out, although its data are so meager that results are suggestive at best. If extraterrestrial impacts trigger all the extinction peaks in the fossil record, then there should be a periodicity in the ages of craters on the Earth's surface as well. And the preliminary finding is just that. It is based on a painfully small number of craters: only thirteen craters known from the past 250 million years have good age determinations. But the periodicity in their ages is about 28 million years—very close to the extinction dates, given the uncertainties in the available dating techniques—and the crater ages cluster right in phase with the extinctions. This is hardly proof, but it is too intriguing to ignore. The search for positive proof has now begun with scanning of the heavens for a tiny solar companion, predicted to be near its maximum distance from the Sun. Perhaps it is there, but right now most astronomers have strong doubts.

I have already mentioned the uneven size of the extinction peaks of the past 250 million years. Assuming that impacts are involved in these extinctions, one explanation for the variations might be that the size of the extinction depends on the conditions on Earth when an impact occurs. Returning to the correlation between regression and major extinctions, we know that, when the sea level drops, the climate deteriorates. Terrestrial and marine environments change, perhaps rendering the Earth's inhabitants particularly vulnerable to the additional stresses that would come if an impact occurred at the same time. On the other hand, if the impact came when sea level was high and climates constant and mild, the biota might be more resilient, and a much more modest extinction would result. This two-factor view of mass extinction is going to be tested in the next year, and even if it does not hold up, it is likely that other complex models for the causes of mass extinctions in the fossil record will soon be proposed.

Evolutionary Effects

Whatever the causes of mass extinctions, their severity and frequency clearly make them a major evolutionary force, one that we still do not fully understand. My own preliminary work (see note 1) and a few hints from previously published studies suggest that mass extinctions involve a change in the rules of extinction and survival, with survival having little relation to the success a group experiences during normal background times. To analyze these evolutionary dynamics I turned not to the spectacular but fragmentary record of the dinosaurs, but to the more prosaic and much richer fossil record of marine clams and snails. Their preservation is so much more complete that the fates of many evolutionary lineages can be traced in some detail before, during, and after the Late Cretaceous extinction.

During background times, lineages containing many species tend to resist extinction. This makes sense: a species-rich lineage can afford the loss of a few species here and there, whereas a lineage with only a few species is much more vulnerable to eradication, even during a period of modest background extinction rates. After all, loss of only two species brings a three-species lineage perilously close to the brink, while it has little effect on a fifty-species lineage. However, this intuitively satisfying rule did not hold for the Late Cretaceous mass extinction: clam and snail lineages rich in species fared no better than species-poor lineages. Among the clams, about 40 percent of both the victimized and surviving lineages were originally species rich. For the snails, about 50 percent of the victims and only about 33 percent of the survivors were species rich. If anything, the species-rich taxa are *under*represented among the survivors.

Larval development also plays an important role in determining extinction and survival during background times. Species and lineages with free-swimming, high-dispersal larvae, which feed on the plankton, last almost twice as long as lineages lacking that combination of traits. But not during the Late Cretaceous extinction. Once again, there was no difference in survival between lineages with different modes of larval development, even though this factor had been important for the 20 million years before the extinction and had regained its importance soon after the mass extinction was over.

Despite this sweeping disregard for the usual rules of survival, all is not anarchy during mass extinctions: New rules take over. The total geographic range of lineages at the time of the extinction played an important part in deciding which lineages survived and which were lost at the end of the Cretaceous. Widespread lineages had a better chance of getting through the extinction than localized ones. For example, among the clams, 33 percent of the victims but only 3 percent of the survivors were lineages restricted to North America. Viewed another way, 55 percent of the widespread genera survived, but only 9 percent of the restricted ones did. For the snails, the exclusively North American lineages contributed 48 percent of the victims but only 11 percent of the survivors. Thus 50 percent of the widespread genera survived, but only 11 percent of the restricted ones did. So far as we can tell, the same effect held on every continent.

The changeover in the rules suggests that whatever causes a mass extinction does not simply increase the number of species that become extinct; it changes the *kinds* of species that become extinct. Groups that were doing fine during normal times will suddenly be in serious trouble if they are not sufficiently widespread. By the same token, other groups can survive, not because they are better adapted in any conventional sense but because they happen to have the right geographic distribution to survive the crises. This is a highly unorthodox view of the way the evolutionary game works. It suggests that in the long run Earth history is not a matter of the steady improvement of the life on this planet, a progressive replacement of one group by a better adapted one. Instead, replacements can come about when a mass extinction, indifferent to adaptations evolved during background times, breaks the hold one group has on particular ecological niches or life zones and leaves the door open for another group that persisted. A lineage can then rise to prominence, not because it is the superior competitor but because it just happened to have the right set of traits to survive the mass extinction.

Our species is probably here for just this reason. The first dinosaurs and the first mammals evolved at about the same time in geologic history. The dinosaurs underwent an exuberant and eminently successful evolutionary diversification, whereas the mammals remained small, nocturnal scramblers in the underbrush for over 160 million years. Only after the Late Cretaceous extinction removed the dinosaurs and other dominant reptiles did the mammals begin to diversify. Within 15 million years there were bats in the sky, whales in the sea, and a host of mammalian herbivores, carnivores, and scavengers covering the land. The fossil record now suggests that many, if not all, major evolutionary turnovers were driven by mass extinctions. Thus mass extinctions now take on a dual significance: destroyers of diverse, well-adapted lineages and deliverers of new groups into dominance and diversification.

The Lessons of the Past

The mass extinctions in the fossil record have compelling implications for the plight of today's wildlife and for the survival of the human species. The fossil record is telling us, first, that major upheavals can and do occur and that such biological crises can be rapid, irreversible, and unpredictable. Once a species is extinct or a network of interacting species falls apart, it is gone forever.

A second message from the fossil record is that the tropics are the Earth's most vulnerable regions. This is bad news because the tropical forests are important for the well-being of us all. Incredibly rich centers of plant and animal life, tropical forests cover less than 7 percent of the Earth's land surface but harbor more than half of the world's species. A patch of forest in the Amazon jungle or Borneo, for example, contains ten times the number of species as in a similarly sized patch of forest in Massachusetts or Montana. Most of those tropical species are scarce, however. They have small geographic ranges, and the fossil record tells us that this makes them especially vulnerable to extinction. Sadly, the findings of modern ecologists bear this out.[13]

In his excellent book *The Primary Source,* Norman Myers estimates that the Earth is losing tropical forests at the rate of at least 200,000 square kilometers per year, mainly as a result of logging and the clearing of forest for agriculture. If this devastation continues, representing an annual loss of about 2 percent of the total tropical forest area, we will have lost at least half a million species, perhaps as many as a million in twenty years. As Myers has documented, this human-caused mass extinction will destroy a staggering wealth of biological resources. The tropics contain over 4000 species of edible fruits and vegetables, and fewer than fifty of these are being used on a large scale today. The other 3950 species could certainly help relieve hunger in the coming decades but only if those species survive to be cultivated.

The tropics are a source of new breeding lines for plants already under cultivation, providing new genes for resistance to diseases and pests. In the last decade alone, interbreeding with wild strains from the tropics has saved crops of cocoa, sugarcane, banana, coffee, and those two staffs of life, rice and corn. Who knows what genetic resources are being lost forever as the forests are destroyed?

One-fourth of all our medicines are derived from tropical plants. At least 1400 tropical plant species contain substances active against cancer, and that is after testing less than one-tenth of the species present. In spite of this, tropical species are being driven into extinction much faster than they can be tested for medicinal use.

Finally, tropical forests may play an important role in regulating global climate, much as wide, shallow seas did in the geologic past. The wholesale clearing of forests that is occurring today will change climate patterns in the United States: rainfall, winter temperatures, timing of first frost—all these factors are vitally important for our own agricultural production. We are dependent on the tropics, which the fossil record tells us are, in many ways, the most delicate communities of all.

In more general terms, the fossil record shows us that, when the rate of extinction strongly outpaces the evolution of new species, as it is doing with a vengeance today, the stage is set for the kinds of extreme changes that I labeled mass extinctions at the beginning of this chapter. Destruction and extinction may be rapid, but recovery is painfully slow. The reefs at the end of the Cretaceous were not unusual in this regard: they were extinguished in only tens to thousands of years, but they took 10 million years to become reestablished. The appearance of tropical reefs required the evolution of new kinds of reef builders, because the Cretaceous builders, the rudist clams, were gone. Perhaps just as chilling is what we have learned about the likely survivors of these crises: not the most highly evolved life forms nor the most useful to our own species, but the most widespread ecological generalists. The survivors would be the weeds and the cockroaches, not the medicinal or nutritional plants.

Our species, then, is on the brink of causing, single-handedly, the worst mass extinction in 65 million years. The very species that provide, or might provide, a rich harvest of medicines, foods, fuels, raw materials, and even climate regulation are being driven into extinction, forever beyond our reach. It is up to us, as beneficiaries of the last major mass extinction, to reverse this trend. And as the other chapters in this book will try to show, there is still hope for this reversal, before many of the species we hold dear—including our own—also go the way of the dinosaur.

Notes

I thank Susan Kidwell and Mari Jensen for encouragement and comments on the manuscript. The research discussed has been supported by the National Science Foundation, Earth Sciences Section. Finally, I am grateful to Ken Mallory and Les Kaufman for inviting me to participate in this symposium and for so patiently waiting for this chapter to materialize.

1. D. Jablonski, "Causes and consequences of mass extinctions: A comparative approach," in Elliot, 1986 (see readings); D. Jablonski, "Background and mass extinctions: The alternation of macroevolutionary regimes," *Science* 231 (1986), 129–133.

2. J. J. Sepkoski, Jr., "Mass extinction in the Phanerozoic oceans: A review," in Silver and Schultz, 1982 (see readings), 283–289. D. M. Raup and J. J. Sepkoski, Jr., "Mass extinctions in the marine fossil record," *Science* 215 (1982), 1501–1503. J. J. Sepkoski, Jr., "A kinetic model of Phanerozoic taxonomic diversity. II. Post-Paleozoic families and mass extinctions," *Paleobiology* 10 (1984), 246–267. J. J. Sepkoski, Jr., "Phanerozoic overview of mass extinction," in Jablonski and Raup, 1986 (see readings). K. W. Flessa et al., "Causes and consequences of extinction," in Jablonski and Raup, 1986 (see readings).

3. G. F. W. Herngreen and A. F. Chlonova, "Cretaceous microfloral provinces," *Pollen et Spores* 23 (1981), 441–555. S. K. Srivistava, "Evolution of Upper Cretaceous phytogeoprovinces and their pollen flora," *Review of Palaeobotany and Palynology* 35 (1981), 155–173. L. J. Hickey, "Changes in the angiosperm flora across the Cretaceous-Tertiary boundary," in Berggren and Van Couvering, 1984 (see readings), 279–313.

4. A. H. Knoll, "Patterns of extinction in the fossil record of vascular plants," in Nitecki, 1984 (see readings), 21–68.

5. W. Alvarez, L. W. Alvarez, F. Asaro, and H. V. Michel, "The end of the Cretaceous: Sharp boundary or gradual transition?" *Science* 223 (1984), 1183–1186.

6. C. B. Officer and C. L. Drake, "Terminal Cretaceous environmental events," *Science* 227 (1985), 1161–1167; D. M. McLean, "Deccan Traps mantle degassing in the terminal Cretaceous marine extinctions," *Cretaceous Research* 6 (1985), 235–259.

7. O. B. Toon, "Sudden changes in atmospheric composition and climate," in Holland and Trendall, 1984 (see readings), 41–61.

8. W. Alvarez, E. G. Kauffman, F. Surlyk, L. W. Alvarez, F. Asaro, and H. V. Michel, "Impact theory of mass extinctions and the invertebrate fossil record," *Science* 223 (1984), 1135–1141.

9. S. M. Stanley, "Marine mass extinctions: A dominant role for temperature," in Nitecki, 1984 (see readings), 69–117; S. M. Stanley, "Mass extinctions in the ocean," *Scientific American* 250 (June 1984), 64–72.

10. D. Jablonski, "Marine regressions and mass extinctions: A test using the modern biota," in Valentine, 1985 (see readings), 335–354.

11. D. M. Raup and J. J. Sepkoski, Jr., "Periodicities of extinctions in the geologic past," *Proceedings of the National Academy of Sciences USA* 81 (1984), 801–805. D. M. Raup and J. J. Sepkoski, Jr., "Periodic extinction of families and genera," *Science* 231 (1986), 833–836. See also D. Jablonski, "Keeping time with mass extinctions," *Paleobiology* 10 (1984), 139–145.

12. R. A. Kerr, "Periodic extinctions and impacts challenged," *Science* 227 (1985), 1451–1453.

13. J. M. Diamond, "Normal extinctions of isolated populations," in Nitecki, 1984 (see readings), 191–246. J. M. Diamond, "Historic extinctions: A Rosetta Stone for understanding prehistoric extinctions," in *Quaternary Extinctions: A Prehistoric Revolution,* P. S. Martin and R. G. Klein, eds. (Tucson, Arizona: University of Arizona Press, 1984), 824–862. D. Simberloff, in Elliott, 1986 (see readings).

Suggested Readings

The mass extinctions field is moving so rapidly that technical journals and symposium volumes are the only way to keep up. In addition to the books listed below, useful resources include the ongoing series of essays by Roger Lewin and Richard A. Kerr in the Research News section of *Science,* by a variety of authors in the News and Views section of *Nature* and in the Current Happenings section of *Paleobiology,* and by Stephen Jay Gould in *Natural History.* Finally, this chapter is largely based on a technical paper (see the first paper in note 1), which contains about 300 references to the scientific literature.

W. A. Berggren and J. A. Van Couvering, eds., *Catastrophes and Earth History.* Princeton, New Jersey: Princeton University Press, 1984.

P. Ehrlich and A. Ehrlich, *Extinction.* New York: Random House, 1981.

D. K. Elliott, ed., *Dynamics of Extinction.* New York: Wiley, 1986.

R. W. Fairbridge and D. Jablonski, eds., *The Encyclopedia of Paleontology*. Stroudsburg, Pennsylvania: Dowden, Hutchinson & Ross, 1979.

H. D. Holland and A. F. Trendall, eds., *Patterns of Change in Earth Evolution*. Berlin: Springer-Verlag, 1984.

N. Myers, *The Primary Source*. New York: W. W. Norton, 1984.

M. H. Nitecki, ed., *Biotic Crises in Ecological and Evolutionary Time*. New York: Academic Press, 1981.

M. H. Nitecki, ed., *Extinctions*. Chicago, Illinois: University of Chicago Press, 1984.

D. Jablonski and D. M. Raup, eds., *Patterns and Processes in the History of Life*. Berlin: Springer-Verlag, 1986.

L. T. Silver and P. H. Schultz, eds., "Geological implications of impacts of larger asteroids and comets on the Earth," *Geological Society of America Special Paper* 190 (1982).

J. W. Valentine, ed., *Phanerozoic Diversity Patterns: Profiles in Macroevolution*. Princeton, New Jersey: Princeton University Press, 1985.

Appendix

The study of extinction in the fossil record has continued at breakneck pace since this chapter was written in 1986. Although some scientists still champion volcanism, evidence continues to accumulate for an asteroid impact at the end of the Cretaceous Period, and the case seems stronger than ever. Many now interpret the Chicxulub structure—180 kilometers in diameter, buried under the northern coast of Yucatan—as the long-sought, end-Cretaceous impact crater, but as of this writing (January 1993) the debate rages on. Intriguing geological hints of impacts at two of the other major mass extinctions, the Late Devonian and Late Triassic, are not nearly as well corroborated but lend support to the possibility that the end-Cretaceous, impact-driven event was not unique. Sea-level and climatic fluctuations still have not been ruled out as contributing or even dominant factors for these and other extinctions, however. In short, there has been a huge influx of intriguing new information bearing on the causes of mass extinctions, but few final conclusions. The present lack of conclusive evidence simply reflects the immense complexity of these questions and the difficulty in reconstructing ancient, perhaps very brief, events that are completely off-scale in terms of modern-day analogs.

Qualitative differences between background and mass extinction have been found in a number of studies. In my own research, end-Cretaceous mollusks from Europe and North Africa conform to the patterns seen in North

America: widespread lineages survived preferentially, but species-richness offered no protection during the mass extinction. Similar patterns are now known for early Paleozoic trilobites, late Devonian corals, Paleozoic cephalopods, and end-Cretaceous sea urchins. Extinction events also preferentially removed, or afforded no advantage to, lineages normally at low risk among Paleozoic bryozoans, late Cenozoic plankton, and late Cenozoic clams. Remarkably, the one reported exception to this general pattern has been for snails during the greatest mass extinction of all, at the end of the Permian; here, species-richness does seem to help during both background and mass extinction.

Overall, however, evidence is accumulating that lineages and adaptations may be lost in mass extinctions not because they were inferior by the standards of background times, but because they lacked the geographic distributions—or other, still poorly understood features—necessary to survive these relatively brief crises. This doesn't mean that the rules of extinction and survival must *reverse* during a mass extinction, but that some of the usual advantages melt away while other factors, previously less important, come to the fore. The role of these events in driving dramatic changeovers in ecological dominance, and in fostering bursts of evolutionary creativity during recovery intervals, is becoming widely accepted.

The lessons of the fossil record for present-day extinctions have been discussed in papers by Jablonski and by Myers, and in Eldredge's recent book (see list that follows). Most of the points raised in this chapter still hold: that the fossil record gives stark evidence that species and communities are not infinitely resilient, so that rapid and irreversible ecological changes can occur; that at least some tropical communities, and geographically restricted species and lineages in general, are particularly vulnerable (I now suspect that not all tropical communities are equally fragile); that recoveries are geologically rapid but immensely slow on human time scales; and that survival may often be indifferent to adaptations honed over millions of years of natural selection.

There is a growing awareness, however, that direct comparisons between the ancient mass extinctions and the present biodiversity crisis can only be pushed so far, given our present state of knowledge of past and living biotas. Uncertainties inevitably arise when we try to make more exact predictions from the fossil record. For example, the extinctions detected by paleontologists mainly involve lineages that are more widespread and abundant (and thus more likely to be fossilized) than the extremely localized species that make up a substantial part of the present-day extinction estimates.

There is little room to doubt that today's unchecked habitat destruction will eventually remove biodiversity on a scale comparable to the major upheavals of the geologic past. But exactly where we are in that process is still unclear—not least because of the crippling lack of an inventory of the world's flora and fauna. It should be obvious that this is in no way support for a "wait and see" approach to the problem.

The fossil record of the past five million years offers special opportunities for predicting upcoming biotic changes. Not only does this interval offer the finest available time resolution, but it also includes a series of glacial and interglacial (warming) cycles that provide natural experiments on how species behave in the face of rapid shifts in global temperature and rainfall patterns. Perhaps the most important message so far is that ecological communities do not respond as units to climate change. Plant and animal species are idiosyncratic in their migration behaviors, so that few, if any, communities existed in their present form a mere 10,000 years ago, at the end of the last ice age. Instead, data from every continent show that communities, over a time scale of centuries to millennia, are really snapshots of continually changing, overlapping distributions, with each species following its own course in terms of abundance and geographic distribution.

Shared habitat preferences, such as favored temperature, humidity, and soil conditions, do play an important role in determining which species occur together, but as these factors change, for example by a decrease in rainfall and increase in summer temperatures in the central United States, different species will respond according to their own tolerances; a simple, *en masse* northward shift of whole communities would not be expected. Some species might like the drier weather but seek higher elevations to keep cool in the summer, while others might like the longer summers but shift east to find more rain, and so on.

Such reshuffling of communities will surely continue in response to normal climate trends and would be accelerated by climatic or other large-scale changes caused by human activities. This basic paleontological insight cannot be ignored in designing nature reserves, or in any other attempt to anticipate the long-term success of populations in a given region, whether crops, livestock, or their pests. Nature reserves, for example, must be sufficiently large and environmentally diverse to accommodate the inevitable migrations following on the heels of climate changes.

To take a simple case, the fossil record shows that reserves should include both high and low elevations, so that species with different temperature requirements can variously adjust their ranges up- and downhill within the limits of the reserve as needed, rather than by larger jumps across latitudes that may be barred to them by human activities. As we carve up natural habitats, migration routes are cut, so that even species mobile enough to cope with most climatic changes are put at risk. And, of course, potential pests can be herded in unpredictable—and undesirable—directions by those changes as well.

Despite all these uncertainties, the fossil record is our only direct source of information on how biological systems respond to large-scale disruption. The past mass extinctions are fascinating, intricate puzzles in their own right and offer profound insights into how our present biological world took shape: why we, and bats, and whales are here, and why dinosaurs, pterosaurs, and mosasaurs are not. There is plenty more to learn, and we just have begun to explore the opportunities and limits of the fossil record as a predictive tool for today's biotic stresses. Comparative research on living and fossil lineages should help us understand, anticipate, and perhaps manage the biological upheavals now driven by human activities.

Additional Readings

D. E. G. Briggs and P. R. Crowther, eds., *Palaeobiology: A Synthesis*. Oxford: Blackwell, 1990 (Mass Extinctions, p. 160–209).

S. K. Donovan, ed., *Mass Extinctions: Processes and Evidence*. New York: Columbia University Press, 1989.

N. Eldredge, *The Miner's Canary: Unraveling the Mysteries of Extinction*. New York: Prentice-Hall Press, 1991.

D. Jablonski, "The biology of mass extinction: A palaeontological view," *Philosophical Transactions of the Royal Society of London* B325 (1987): 357–368 (mostly comparing background and mass extinction).

D. Jablonski, "Extinctions: A paleontological perspective," *Science* 253 (1991): 754–757 (part of a special issue on biodiversity).

G. P. Larwood, ed., *Extinction and Survival in the Fossil Record*. Systematics Association Special Volume 34. Oxford: Clarendon Press 1988.

N. Myers, "Mass extinctions: What can the past tell us about the present and future?" *Palaeogeography, Palaeoclimatology, Palaeoecology* 82 (1990): 175–185 (part of a workshop on global change).

D. M. Raup, *Extinction: Bad Genes or Bad Luck?* New York: W. W. Norton, 1991

V. L. Sharpton and P. D. Ward, eds., "Global catastrophes in Earth history," *Geological Society of America Special Paper* 247, 1990.

3

The Amazon: Paradise Lost?

Ghillean T. Prance

The Amazon Environment

The Amazon Basin (figure 3.1), with its vast expanse of forest, is the world's largest tropical rain forest. The forested lowland parts alone cover an area equal to the United States east of the Rockies. If separated as a nation, the basin, or Amazonia as it is known to the Brazilians, would be the world's ninth largest nation, encompassing nearly 60 percent of Brazil, or over 1.235 billion acres of land. The Amazon River is responsible for the drainage into the sea of one-fifth of all the freshwater in the world. It is also biologically the most species-rich area of the world, containing over 50,000 species of higher plants, at least an equal number of fungi, a fifth of all the birds on Earth, at least 3000 species of fishes, amounting to ten times the number of fish species in all the rivers of Europe, and insect species numbering in the uncounted millions.

The Amazon forest contains a vast array of fascinating and record-beating plants and insects. The world's largest snake, the anaconda, graces the river banks of Amazonia, and one of the world's largest insects, the rhinoceros beetle, flies through its forests. A host of useful products of worldwide importance have come from the forests of Amazonia, including cacao (the source of chocolate), rubber, quinine, Brazil nuts, and chicle, or chewing gum.

Between the years 1600 and 1900 mankind eliminated about seventy-five known species worldwide, almost all of them mammals and birds.[1] During the present century, however, extinction has increased at an alarming pace, especially in the tropical rain forest. There are many causes of tropical extinctions, but they are all related to the extensive local endemism of species. And destruction is now on such a large scale in certain regions that it covers more than the entire range of these endemic or localized species. The Volkswagen Motor Company, for example, holds a concession

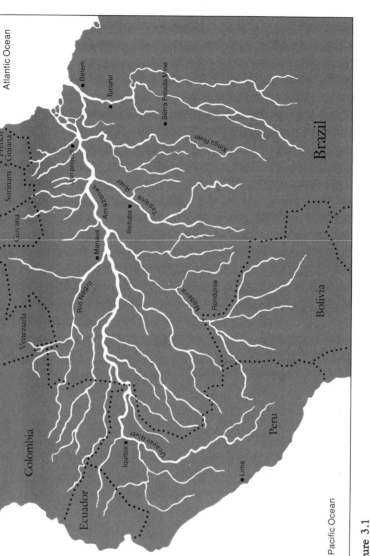

Figure 3.1
The vast network of the Amazon Basin. Map prepared by David Vergara, of the New England Aquarium.

of 346,000 acres (1400 square kilometers) of forest. It is turning much of this area, half the size of the state of Rhode Island, into cattle pasture. The Jari forestry project, founded by Daniel K. Ludwig, covers 2.5 million acres—larger than the island of Puerto Rico or the island of Crete or Cyprus. Major tributaries of the Amazon, such as the Tocantins River, are being dammed to produce hydroelectricity, completely disrupting the life of fishes and other aquatic organisms, and flooding vast areas of riverine forest. Thus we have cause for concern about extinction in Amazonia, a region where not just individual species but whole habitats are disappearing rapidly.

The Amazon Paradise

To refer to the Amazon as a "paradise lost" assumes that the region, which has more generally been called a "green hell," was at least at one time a paradise. For many pioneers and early explorers, who fought diseases, battled with Indians, and were overwhelmed by the quick regrowth of the forest, it must have seemed like hell. In 1971 archaeologist Betty Meggers ameliorated the judgment of "green hell" by calling Amazonia "a counterfeit paradise" based on her studies of the relationship between dwellers and the region.[2] To a biologist, the Amazon *is* a paradise but one precariously balanced between being regained or lost forever.

The depths of the tropical rain forest are awesome to enter. A sense of quiet dignity pervades the great interior, where one is surrounded by massive trunks rising pillarlike to the vaulted arches of branches and the green ceiling of layered leaves. There is a stillness and a quiet made almost palpable by the contrast of the sudden, strange whistling call of the screaming piha bird (*Lipaugus vociferans*). The damp, decaying leaves muffle the sound of footsteps, and only the snapping of a twig or whine of an insect breaks into the solemn serenity.

The plant life takes many forms that are quite unfamiliar to inhabitants of the temperate zone (figure 3.2): trees are huge and buttressed, or slender and smooth barked; climbing vines called lianas are squared, rounded, or crenellated; a host of epiphytic bromeliads and orchids live attached to trees; and there are myriad fungi, miniature, delicate organisms that are nourished by the rotting leaves and are critical to the recycling of the forest.

I first described the paradise, then talk about its impending loss, and finally outline the remaining possibilities of regaining the paradise.

It is only after years of experience, as we learn something about its workings, that the forest begins to emerge as a world of wondrous intricacy.

a

Figure 3.2
Three components of the rain forest community. (*a*) An open forest on white sand displays its many bromeliads. (*b*) A liana, or vine, is shown wrapped around a rain forest tree in French Guiana. (*c*) Many vines share stems of unusual structure, as in this *Bauhinia* vine. Photographs by Ghillean Prance.

b

c

As one understands more about the species diversity, the complex interactions between the different organisms, and the number of useful plant and animal products that the forest contains, the initial feeling of worshipful awe begins to make sense.

Study of Amazonia is now revealing that the rain forest is a delicately balanced and fragile ecosystem in which each component is dependent on many others. But this study is also humbling because of our genuine lack of knowledge and the surging questions for which we have no answers. We have learned, for example, that regional rainfall is largely controlled by the forest itself (figure 3.3), because over 50 percent of this rain originates from the transpiration of the trees,[3,4] and yet, before we have

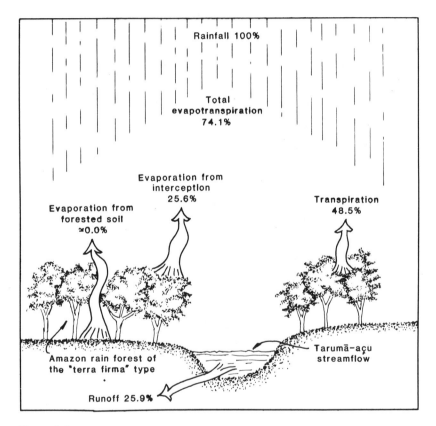

Figure 3.3
Water balance from a study of a model basin near Manaus, Brazil. From E. Salati and P. B. Vose, "Amazon Basin: A system in equilibrium," *Science* 225 (1984), 130. Reprinted by permission.

even named the tree species or studied the interactions of trees and climate in any detail, commercial development is subjecting the forest to unprecedented devastation.

Let us examine the intricacy of the undisturbed forest—before the massive interference that is occurring today—that led many naturalists to consider it a paradise. Despite the hardships that he suffered in Amazonia, Henry Walter Bates ended his famous book on the Amazon by calling the region a "Naturalist's Paradise." Yet even this great naturalist expressed reservations about the true nature of the region:

> But now, after three years of renewed experience of England, I find how incomparably superior is civilized life, where feelings, tastes, and intellect find abundant nourishment, to the spiritual sterility of half-savage existence, even though it be passed in the garden of Eden.[5]

Perhaps these feelings are re-expressed today in the conflict between preservation of the Amazon paradise and its destruction to gratify the immediate demands of modern civilization.

The Amazon Rain Forest: Wonders and Weaknesses

Species Diversity
Without doubt the most important reason that we should be concerned about the destruction of Amazonian forests is because of the vast number of species they contain. The threat of extinction is so great in the Amazon because so many species live as specialists adapted to particular habitats and because so many exist in a limited range.

In the Amazon region there is a rich variety of species of each general type of organism that is unmatched in other places in the world. Henry Bates, who spent eleven years in the Amazon region, from 1848 to 1859, described 8000 new species from his 14,712 collections.[6] Today, only a small percentage of the insects have been named. A single hectare (2.47 acres) of forest that I examined in detail near Manaus, Brazil, contained 179 species of trees of 15-centimeter diameter or more and 236 species of 5-centimeter diameter or more.[7] The record for tree diversity is at Yanomono in Amazonian Peru where Alwyn Gentry found 300 species of trees and lianas in a single hectare of rain forest. Compare this with the five or six species that might be found in a similar area of New England forest and the difference in structure and composition between the temperate and tropical forests becomes obvious. Many similar studies have shown that the species diversity of forest trees is high throughout the region, varying a bit according to the local rainfall.[8]

Many factors cause this diversity of species, and several conflicting theories have been proposed to account for it. Earlier workers thought that the stability of the forest over many years caused its diversity by allowing many species to evolve and accumulate.[9] A more recent view has pointed out considerable fluctuation in the area of rain forest as the world climate changed. There was less forest, for example, in the colder, drier climates of the Pleistocene glaciations. This has led some people to conclude that it was not stability, but instability of the forest over time that led to the diversity.[10–13] A different kind of hypothesis was proposed by Janzen.[14] He suggested that a continuously warm climate favored the development of many types of herbivorous insects that feed on specific plant species. Thus the diversity of plants could be a secondary phenomenon—that result of selection that favored differences that allowed plants to escape each specific insect predator. A consequence of this secondary diversity of plant species is that individuals of particular species may be scattered and therefore harder for a particular herbivorous insect to find. Thus spatial heterogeneity of plant species and mechanisms to enhance it (such as increased seed dispersal) would also be favored by natural selection.

Habitat Diversity

Some ecologists have favored the hypothesis that the number of niches available in tropical rain forests has led to species diversity.[15,16] This certainly has contributed to the diversity because there are so many habitats available to the plant species. One can still fly a small plane over areas of undisturbed Amazon forest for several hours at a time, and it appears to be a uniform mass of tall forest. Only a few differences are noticeable, such as the various shapes of the crowns of the trees and the difference between the canopy level trees and the taller emergents, which rise above the canopy. Studies of the vegetation, however, have shown that there are many types of forest that comprise this seemingly uniform Amazon forest.

The greatest expanse is the upland nonflooded forest on terra firme. Species content and structure is quite different in forest that is inundated by the annual rise of the rivers (up to 45 feet in some places) compared with forest on terra firme. And terra firme forest on slopes is often rather different from forest on level parts. In some areas, particularly south of the Amazon between its tributaries, the Tapajós and Xingu rivers, there is much liana forest with an exceptionally high frequency of woody vines that envelope all the trees.

Forest flooded by the so-called white waters or muddy rivers (várzea forest) is quite different from forest flooded by the black waters (igapó forest), which lack sediment but are stained like black coffee because of the suspended humic substances from plants.[17]

Forest in areas where there is white sand soil, such as in parts of the Guianas and in the region of the upper Rio Negro in Brazil, have yet another forest type with many different species and often a more tortuous growth form. In Brazil the white sand forest is termed *caatinga*. There are many endemic species in the caatingas of the Rio Negro and adjacent Venezuela.

Not even these major divisions of vegetation present a uniformity within themselves, however. Species in tall forest on terra firme can find many different niches. Some species are emergent trees that rise above the canopy, such as the Brazil nut (*Bertholletia excelsa*) or the largest tree of the Amazon forest, the majestic legume *Dinizia excelsa*. Some trees in this stratum of the forest have seeds that are adapted for wind dispersal because the wind reaches these trees. The genera *Cariniana* and *Couratari* in the Brazil nut family, for example, are the only two from that family with winged, maplelike seeds. Either they are tall emergents in the terra firme forests or they grow along open riversides or in open windswept savannas. Below the emergent species are those that form the thick canopy that shades all other species beneath. Yet other species are confined to the subcanopy levels and never grow up into the canopy. Further species are low treelets and shrubs at eye level as one walks through the forest. Finally there is the ground cover of small herbs and seedlings.

Plants that grow on the forest floor must adapt to low light because often only 2.5 percent of the incident light above the canopy reaches the forest floor. Many rain forest plants, such as *Philodendron, Monstera,* and *Calathea* have become popular houseplants because of their low light requirements. Accompanying the group of plants at each level in the forest are the animals and insects that interact with them. One group of insect pollinators forages in the lower levels, another in the canopy. These few facts show that there are indeed many niches in the Amazon forest and that this contributes to its diversity. Indeed diversity begets diversity. Moreover, I have not even touched on the quantity of epiphytes or air plants perched on the tree branches or on the large number of nonforest vegetation types that occur within Amazonia, including savanna and open white sand *campina*.

Relationships between Species Dwelling in the Rain Forest

Although the diversity of habitat and species within the Amazon forest makes it seem like a paradise to the naturalist, it is the closely coevolved interactions between the various organisms that are the true marvel. Such interactions are not absent in the temperate forest, but they are simply not as numerous or bizarre as they are in the tropical rain forest. There are, for example, the many ways in which plants are pollinated or get their seeds dispersed by animals or use insects to defend them against other insect predators. There are nonpoisonous insects that mimic poisonous ones for protection. There are insects that perfectly match green lianas on trees, dead leaves on the ground, twigs of plants, or the bark of a tree. The early nineteenth-century naturalists Alfred Russel Wallace, Richard Spruce, and Henry Bates described some of these interactions. It was Bates who described the phenomenon of mimicry in butterflies for the first time. Every organism interacts with another in the ecosystem so that the species are linked through this complex web. When the forest is disturbed, this web is unraveled, and species are lost.

The Pollination of the Royal Water Lily

The royal water lily (*Victoria amazonica*; figure 3.4) is one of the best-known plants of Amazonia. It has magnificent floating pads up to 8 feet in diameter, with turned-up margins showing the spines that protect the pads from aquatic predators. The lily produces a large white flower that opens at dusk. It is an unforgettable sight to watch the mass of white stars open as darkness falls. The white flowers are up to 9 degrees centigrade warmer than air temperature, and they are strongly scented, smelling rather like an overripe pineapple.

As soon as the flowers open, large brown scarab beetles (*Cyclocephala castanea* and *cyclocephala hardyi*) begin to arrive. They enter the large cavity in the center of the flowers and begin to feed on starch-rich food appendages at the top of the cavity. Later in the night the flowers close up tightly and trap the beetles inside. The beetles are content with food and a warm resting place, and they do not damage the vital parts of the flower because they provide food. The flower remains closed during the next day and opens again at night. The flower looks different now, for it has changed color completely, and when it reopens on the second night, it is dark purple and red. It is no longer scented or heated above air temperature. As the flower opens on the second evening, it releases its pollen so that the pollen adheres to the beetles as they emerge, sticky with plant juices from inside the cavity. The beetles then fly off into the night to find a white flower to enter. The stigma of the white flower is receptive and ready to receive

Figure 3.4
(*a*) Visible here are the flowers and leaves of the royal water lily (*Victoria amazonica*); the bud at the lower right is ready to open at sunset. (*b*) A nocturnal scarab beetle (*Cyclocephala castanea*) pollinating the flower of an Amazonian water lily. (*Nymphaea amazonum*). Photographs by Ghillean Prance.

pollen, and this is the way the beetle transports pollen from plant to plant. Because a plant produces a new flower only every second day, the beetles ensure that the pollen is carried to a different plant, so there is cross-pollination.

The Brazil Nut (*Bertholletia excelsa*)   The Brazil nut (figure 3.5) has a complicated flower structure. The stamens around a ring at the center of the flower are protected by a tightly closed hood over the flower. Nectar is produced inside the hood at the base of a large number of staminodes. The hood, which protects the flower parts from ready access by small insects, can be lifted only by a large, strong insect that can push up against the springlike action of the ligule that attaches the hood to the basal ring of stamens. Our work[18] has shown that only large bees, such as the carpenter bee (*Xylocopa*), large bumble bees (*Bombus* sp.), and the orchid bees (*Eulaema* and other large Euglossinae), can lift up the hood. The bee lands on the hood's top and finds its way into the crack to forage for nectar in the hood. At the same time the pressure of the spring forces the back of the bee hard against the pollen-producing stamens and the stigmatic surface in the center of the flower. The bee is then dusted with pollen and will transport it to another flower when it continues its foraging route.

Brazil nuts are produced in a large, round, hard, woody fruit about twice the size of a baseball. The fruits, which take fourteen months to develop after pollination, fall to the ground in January and February during the rainy season. The fruit case is so hard that it can remain for several years without rotting, whereas the seeds or nuts, arranged inside like the segments of an orange, would lose their viability if exposed. A large rodent, the agouti (*Dasyprocta* sp.), chews open the outer shell of the fruit and removes the nuts. It then buries caches of Brazil nuts away from the tree. Because the agouti does not find all its caches, the nuts are dispersed around the forest. The production of one of the most valuable products of the Amazon forest is therefore dependent on large bees that visit the top of the tree above the canopy and on a rodent that forages on the forest floor.

The Sapucaia (*Lecythis pisonis*)   The sapucaia, a relative of the Brazil nut tree, also has a large, woody fruit. It is larger than the Brazil nut and is bell shaped. When the fruit is mature, only a circular lid (operculum) falls off, leaving a large "bell" with the seeds hanging in the middle on long, fleshy stalks, like a whole row of clappers in the center of the bell. The cavity left by the detachment of the operculum is large enough to allow bats access to the seeds. The bats eat the fleshy stalk of the seed and so, at the same

Figure 3.5
(*a*) The fruit of the Brazil nut (*Bertholletia excelsa*); the nuts are arranged much like the segments of an orange inside a large woody case. Drawing courtesy of Gray Herbarium library, Harvard University. (*b*) A nut gatherer uses his machete to crack the hard outer shell and to get to the meat. Photograph by Ghillean Prance.

time, transport the seeds across the forest. It is common to find a collection of uneaten sapucaia seeds under a bat roost. This tree uses a flying mammal to transport its seeds, unlike its relative the Brazil nut tree, which uses a terrestrial mammal.

Ant Plants Several unrelated genera of plants, *Cordia* (Boraginaceae), *Duroia* (Rubiaceae), *Hirtella* (Chrysobalanaceae), and *Tococa* (Melastomatacae), have developed pouches at the base of their leaves or on the leaf stalk. Ants inhabit these pouches. All these plants have long hispid hairs, and the small ants can travel between the hairs, thus protected from other insects. Ants protect the plant from other insect predators. There are many antinhabited plants in the Amazon forest (figure 3.6), and in most cases the plants produce food for the ants as well as provide a home. Such trees as *Tachigali* (Caesalpiniaceae), *Cecropia* (Cecropiaceae), and *Triplaris* (Polygonaceae) are inhabited by ferocious fire ants that offer protection to the tree.

Other trees produce extra floral nectaries to feed the ants that protect their leaves. *Inga* (Mimosaceae) is a good example because there is a row of nectaries between each pair of leaflets along the rachis of its compound leaf. The nectaries reach their full size in the leaf bud before the leaflets fully unfold, and thus the young, tender leaves are well guarded by the ants that visit these nectaries. The andiroba tree (*Carapa guianensis*), the source of a valuable medicinal oil and of good timber, is yet another example of the many young fruits of the Amazon forest that has extrafloral nectaries and protective ants.

A Bird–Insect Relationship Many travelers have observed the curiosity of a wasp or hornet's nest close to a colony of the cacique birds (*Cacicus* spp.), which weave long, pendulous, basketlike nests (figure 3.7). The birds choose to build near the hornets because these insects drive off botflies. The botflies' larvae can threaten the cacique chicks. Nature has developed for the chicks a protective relationship with an insect we commonly regard as noxious. The story is not as simple as that, however, because the cowbird, like the cuckoo in Europe, lays its eggs in cacique nests. Neal Smith, a zoologist from the Smithsonian Tropical Research Institute in Panama, noticed that some colonies of caciques reject the cowbird's egg, but others allow the parasite chick to hatch and grow. Smith eventually found that caciques nesting with the hornets throw the cowbird eggs out, whereas those without the insect protectors rear the cowbird chicks. He then found that cacique chicks in a nest with a cowbird chick are nine times less likely to have botflies than those growing up without cowbirds or hornets. The

Figure 3.6
A young calabash fruit (*Crescentia cujete*) with a protective coating of ants on its extrafloral nectaries. Photograph by Ghillean Prance.

Figure 3.7
A cacique bird (*Cacicus* sp.) and its nest, a common sight along Amazonian rivers.
Photograph by Loren McIntyre.

reason is that cowbird chicks soon open their eyes and are much more aggressive than the cacique chicks. They defend themselves and the cacique chicks from botflies, and we thus have a complex pattern of behavior involving two species of bird and several of hornets and wasps.

The Leaf Cutter Ant   A common sight on the floor of the Amazon forest is a trail of leaf fragments moving along the ground. Closer examination shows that the pieces of leaves are moving because they are carried by ants (figure 3.8), which have snipped them off nearby trees. The ants transport the leaves to gigantic, complex underground chambers that they inoculate with a certain species of fungus. The growing mass of fungal mycelia provides food for the ants, thus making the ants miniature underground

Figure 3.8
Leaf cutter ants carrying their harvest to their underground nest.

gardeners. The fungus is just a mass of unidentifiable, threadlike hyphae. Researchers at the New York Botanical Garden have grown these fungi in glass jars or culture media to make them produce the mushroom so that they can be correctly identified.

Observing these ants led to an unusual observation. If one places a random selection of forest leaves on the trail, the ants will chew up some of these leaves and take them to their nests, whereas they will simply clear others off the trail and reject them as trash. Some of the rejected leaves have been found to contain natural fungicides, and thus one of the best ways of discovering new fungicides is to observe the behavior of the leaf cutter ant. Its delicate chemical receptors, which can detect substances that would destroy its food garden, give clues about which plants contain these natural fungicides.

The Sloth (*Bradypus tridactylus*) The three-toed sloth is not just a slow-moving, lazy mammal; it is a mobile, multi-organism ecosystem. Well camouflaged because of the green algae that live on its gray fur, the sloth is also host to beetles, ticks, and mites that live protected in its fur. The most interesting hitchhikers are the species of moths. The sloth descends from the trees to defecate only occasionally (once in three weeks). One species of moth lays its eggs in the sloth dung where the larvae develop until the mature moths link up with the sloth's next ground visit.[19]

I have touched on only a few interactions out of myriad possible choices. The examples given, however, show that the undisturbed forest has a wondrous intricacy—it is indeed a paradise that should command our reverence and respect. To tamper with one species automatically causes problems for numerous other species that interact with it. It is this extreme degree of interconnectedness combined with the great number of species that make human interference so serious a problem in the tropical rain forest.

Peoples of the Amazon   For at least the last ten thousand years a large number of indigenous peoples, the Amazonian Indians (figure 3.9), have inhabited the Amazon region. By the time Europeans reached the Amazon in the early sixteenth century, there was a population of at least five million Indians speaking as many as 300 different dialects derived from four major stem languages: Tuppi, Arawak, Carib, and Pano-Tacanon. Today the remnant Indian population is only about a quarter of a million.

Although at the time of the arrival of white settlers the Amazon region sustained a population much larger than it is today, the region was largely a forest with an abundance of species. In other words this population had minimal impact on the environment. Early explorers recounted the abundance of manatees, turtles, and caiman. Their description of the quantity of these animals along the rivers is quite unlike that of today, which is that the animals are rare and difficult to see. Today one could not echo the words of Henry Bates describing the lake margins of the Rio Japurá region in 1850: "This inhospitable tract of country . . . contains in its midst an endless number of pools and lakes tenanted by multitudes of turtles, fishes, alligators, and water serpents" (see note 5, p. 315). Yet we know that the Indians hunted these species as sources of food, oil, and many other products. Today, the manatee, or sea cow, is one of the rarest and most endangered of all Amazon species.

The Indians did cut and burn forest for their plantations of cassava, corn, and other crops. They cut small patches that were surrounded by forest, however, so when the field was abandoned, the seeds were available for the forest to regrow, and the soil microorganisms were also nearby to recolonize the soil. In other words, the Amazon Indian agriculture, which developed gradually over several millennia, preserved the soils, the wildlife, and the ecosystem as a whole.

Today scientists are studying the ecology of the Amazonian Indians so that we might better understand how they have managed not only to survive but also to prosper in the Amazon forest. The work of ethnobi-

a

b

Figure 3.9
(*a*) A Txukahemei Indian in a ceremonial headdress. Photograph © 1984 by the
Cousteau Society, Inc., a nonprofit environmental research organization located at
930 West 21st Street, Norfolk, Virginia 23517. Reprinted by permission. (*b*) A
Jarawara Indian preparing tobacco by drying the leaves over an inverted bowl; the
leaves are then pulverized into a snuff. Photograph by Ghillean Prance.

ologist Darell A. Posey[19,20] has shown us much about the ecology of the Kayapo Indians of the village of Gorotire. The Kayapo practice the traditional swidden agriculture, but they also manage areas that appear to the uninitiated as virgin forest. For example, they actually plant fruit trees along the trails that they use most frequently. These plants are therefore accessible as food on long journeys and are easy to manage as they travel. The work of Christine Padoch among the Bora Indians of Amazonian Peru has shown the extensive use that the Bora make of their "abandoned" fields. The fields for their primary crop were thought to be abandoned after a few years of use. Padoch's research has shown, however, that Indians often control the succession of regeneration so that a large quantity of useful plants is allowed to grow. Over the years the Bora make frequent visits to these old fields to harvest fruits, medicinal plants, edible mushrooms, fibers, fish poisons, and many other products.

The indigenous peoples are an integral part of the Amazonian paradise, a part that has much to teach people from modern industrialized society.

Economic Products of Amazonia

God also said, "I give you all plants that bear seed everywhere on earth, and every tree bearing fruit which yields seed: they shall be yours for food. All green plants I give for food to the wild animals, to all the birds in heaven." (*Genesis I:29–30*)

Just as the Garden of Eden was given to Adam and Eve to use, the Amazon comprises a wealth of useful species that we cannot ignore. One of the reasons most frequently given for the conservation of the Amazon forest is that such a wealth of species also has the potential to yield a wealth of medicinal, food, and fuel plants. My own studies of economic botany in the region certainly indicate that this is so.

Brian M. Boom, of the New York Botanical Garden Institute of Economic Botany, carried out a study with the Chácobo Indians of Bolivia.[21] A hectare of forest in the Chácobo territory was inventoried as in a conventional forest inventory; every tree over 10 centimeters in diameter was collected and identified. Subplots were made to study vines, shrubs, herbs, and all other plants. The Indians were then asked about their names and uses for the plants. Of the 91 species of trees the Indians used 75 of them, 85 percent, in some way. Counting the individual trees standing on a hectare, the Chácobo used a phenomenal 95 percent (619 of the 649). This study shows for the first time the actual value of the Amazon forest to the Indians. It shows that they have discovered a use for the majority of the plant species that grow around them. This further supports statements that the forest has great economic potential. A selected list of the Chácobo's medicinal plants is given in table 3.1. I could give a more lengthy catalog of useful plants of Amazonia, but the Chácobo study adequately illustrates the point.

I have already mentioned the Brazil nut in connection with its pollination. Just two further examples of Amazonian plants of great economic potential will illustrate the value of the plant resources of the region.

The guaraná plant (*Paullinia cupana* variety *sorbilis*) produces the most popular soda in Brazil. It is a plant that is used by the Indians as a stimulant to remove hunger pains and to enable greater work because of its high caffeine content (greater than coffee), and it is now the cola of Brazil. The guarana is made from a vine in the soapwort family, the Sapindaceae. This important Amazonian crop grows well on the poor soils of terra firme and is a much better use of the land than the pastures that have been created over large areas in the past few years. Guaraná remains little known outside

Table 3.1.
Selected Chácobo medicinal plants

Chácobo name	Scientific name	Family	Use	Plant part
bahuaquexti	*Peschiera benthamiana*	Apocynaceae	skin lesions	latex
ahuaranihi	*Psychotria lupulina*	Rubiaceae	diarrhea	leaves
bimichexe	*Psychotria prunifolia*	Rubiaceae	infections	leaves
paxacaxachi	*Diplaria karatifolia*	Cyperaceae	rheumatism	leaves
nibosa	*Piper darienense*	Piperaceae	toothache	root
cashixopame	*Cissampelos andromorpha*	Menispermaceae	toothace	leaves
maxejoni	*Rourea camptoneura*	Connaraceae	rheumatism	fruits, leaves
behisiti corihua	*Protium unifoliolatum*	Burseraceae	infections	resin
pexcanishi	*Davilla nitida*	Dilleniaceae	stomachache	leaves
mamamaxe	*Aparisthmium cordatum*	Euphorbiaceae	stomachache	leaves

Data from B. Boom (see note 21), based on a study conducted at Alto Ivon, Beni, Bolivia, from November 1983 to April 1984.

Brazil and has only recently reached the United States in a limited market; it is one of the many examples of Amazonian crops that could be developed further.

The pataua palm (*Jessenia bataua*) has long been valued by the Indians of Amazonia. This palm has been used for food, fiber, construction materials, medicine, weapons, and toys. The small dark-purple, olivelike fruit is the most important product of the pataua. When it is steamed and macerated, it yields an oil whose physical and chemical properties are identical with olive oil. The pulp that remains after extraction of the oil has a protein value nearly equivalent to milk. The Indians mix this pulp with water and make a nutritious, high protein beverage. This is another vastly underexploited resource of the Amazon forest.

This paradise, the Amazon, has many, many plants that can be utilized if they are not lost.

Paradise Lost: The Destruction of the Rain Forests

O miserable of Happie! is this the end of this new glorious world?
(*John Milton*, Paradise Lost, *Book X, line 720*)

Habitat Destruction

Now that I have sketched an outline of our paradise, I briefly recount some of the things that are happening to it today. The change really began in 1970 when President Medici of Brazil visited the drought and poverty stricken northeast of his country. He was so impressed by the poor estate of his subjects that for the best of humanitarian reasons he decided that the only solution was to develop the Amazon region and thus to open a large area of settlement for the northeasterners. They would be moved in great number from their arid land to a rain-filled paradise, the Amazon. The horror-stricken President Medici made a moving appeal to the Brazilian Congress on June 6, 1970. Only ten days later a National Integration Plan (PIN) was formulated; it provided for the construction of the Transamazon highway, the first leg of which was to cut across the Amazon region from east to west about 500 miles south of the Amazon River. The plan also included an elaborate colonization plan along the length of the highway. Bids for the highway construction were already out by June 18, and the construction began on September 1, 1970, an indication of the priority and urgency of the president's plan. This was the beginning of a particularly destructive phase in the history of the Amazon region (figure 3.10). At the same time, President Fernando Belaunde Terry of Peru, who had been

Figure 3.10
Bulldozing the rain forest. Photograph by Loren McIntyre.

dreaming for many years of building a north-south road around the western fringe of the Amazon, was designing plans to develop the Peruvian Amazon.

The Transamazon highway and many other Amazonian highways (figure 3.11) have now been built and for the most part are functional, but the colonization plan has been a failure. Its plan to relocate one million northeasterners actually moved fewer than twenty thousand people. Even if it had moved the one million as planned, that is less than the annual population increase of northeastern Brazil.

The plan was unsuccessful largely because of poor soils, the failure of the initial rice harvests, and inadequate agricultural advice to the settlers. Many people ended up returning to their area of origin. Only a few of the settlers have succeeded: some by luck, some by hard work, and some by having enough local knowledge to have chosen one of the few areas along

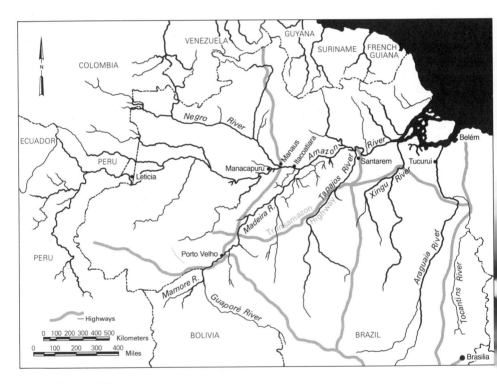

Figure 3.11
This map of the principal road systems of Amazonia shows the network that allows easy access to the forest and has precipitated colonization and deforestation. Map by Nigel Smith.

the highway that is more suitable for agriculture. Today, as one drives along the highway, one is most impressed by the quantity of abandoned land full of regenerating *Cecropia* trees and of degenerated, weed-filled cattle pastures.

Because the colonization plan for the highway failed, the next president of Brazil, Ernesto Geisel, concluded that the Amazon was not suitable for colonization by the small farmer and that capital-intensive development was needed. Accordingly, he created huge tax incentives for companies and rich individuals to invest in enormous tracts of land. The development passed from the phase of *minifundarios* to that of *latifundarios*. Vast areas of forest were felled, burned, or treated with herbicide to be replaced with cattle pasture in a region that is not suitable for cattle raising. Many of the areas cleared for such ranches have now been abandoned. Frequently the owners did not care if they failed because the tax incentives and speculation in land values ensured that they would have no economic loss. In many areas it takes about 1.5 hectares or more to sustain a single cow. This cow lives on an area that could have had over 700 individual trees of about 200 species, and many other plant and animal species as well—an enormous natural biomass with much greater productivity and value than that offered by the skinny, malnourished cow that now wanders around the weed-infested pastures that have replaced the forest. The failure of many ranches has led to a slowdown in the rate of deforestation for that purpose.

The state of Rondônia, on the western frontier of Brazil, provides another alarming chapter in the destruction of Amazonia. This area, which for a long time experienced only a moderate disturbance from tin mining and the construction of the famous Madeira-Mamoré railroad, has now become the development frontier of the Amazon region. When it was discovered in 1974 that, unlike the region of the Transamazon highway, there are rather large areas of the rich terra roxa soils and that the terrain is flatter and more suitable for agriculture, what can only be described as a "land rush" of enormous proportions began along highway BR-364 between Cuiabá, in Mato Grosso state, and Porto Velho, the booming capital of Rondônia. The colonization began to explode during the 1970s, and the government colonization agency, INCRA, was overwhelmed by the quantity of new settlers arriving in Rondônia. Soon entrepreneurs, such as the famous Melhorança brothers, much discussed in the Brazilian press, set up businesses in land speculation that rivaled the work of INCRA. As the government caught up with speculators, many settlers found that they had no title to the land that they held and into which they had put their meager life savings. In a region of rain forest that is suddenly invaded in this way, people take law into their own hands, and the gun becomes the law.

Shootouts are common, such as one I witnessed in a bar in Vilhena, the entry town to Rondônia, which left two persons dead. The population of Rondônia has increased by over a thousand percent since 1968. The result is a devastation of the forest.[22]

Today's aerial photos of the region compared with those of a few years ago show a vast network of new roads intersecting BR-364 and enormous areas of decimated forest. The only difference between BR-364 and the Transamazon highway is that in the former case many of the resulting farms work, for two reasons. First, the soil is better, and second, a diversity of crops has been grown. With less government control, success was not based on a single crop, rice, which was emphasized in the Transamazon colonization. The diversity of crops that the Rondônia settlers used has meant that, although some failed, others succeeded. Often when one farmer experienced a crop failure, another member of the family had a successful crop on his land and carried his relatives through the hard times. In addition, many settlers in Rondônia have concentrated on woody perennial crops, such as cacao, the chocolate plant (*Theobroma cacau*), which is less destructive to the soil than most annuals. Consequently, the settlement of Rondônia is a curious mixture of unplanned deforestation and species loss and an interesting and often successful experimental laboratory in ways of using the land.

But the real tragedy of Rondônia is that it was one of the most important centers of species diversity in Amazonia. The same reasons that make the region useful, such as its richer soils, also made it the home of an extraordinarily high number of endemic species of plants and animals.

Rondônia is thought to have remained forest covered during the drier periods of the Pleistocene glaciations, thus providing a haven for many forest species. Such forest refugia are the most important places to conserve the genetic diversity of Amazonia. The one national park in Rondônia, Jau, covers only a small part of the Rondônia refugium, and its borders have already been invaded. Another part of the Rondônia refugium has been flooded by the lake behind the Samuel hydroelectric dam. Less uncontrolled but equally devastating deforestation is taking place further west, in the state of Acre, the new frontier; the story promises to be the same.

The Amazon forest is disappearing for many other reasons, but the rush for land and agricultural settlement are certainly the most immediately destructive and the most alarming. Other projects that are changing the region from paradise to a devastated fragment of its former splendor include forestry projects, hydroelectric dams, and mining enterprises.

The vast Tucuruí dam, slated to produce 8000 megawatts of electricity, was closed in 1984 and has created an enormous 100-mile-long lake on the Tocantins River. Much research was carried out in the area of the lake before the dam was closed. The commendable scientific work of the Museu Goeldi in Belém and the National Amazon Research Institute (INPA) in Manaus has meant that much was learned about this region before the flood. Fortunately, it is not one of the centers of biological diversity, so that loss of terrestrial species there will not be great. Even so, the dam has caused many problems. The efforts to remove the standing crop of timber from the area to be flooded was a series of disasters that led to the bankruptcy of the military pension scheme. Despite contracts to both the pension scheme and a French company for removal of the timber, the timber was never cleared, and the lake has flooded much of the forest. The company, CAPEMI, had no logging experience and went bankrupt after clearing only one-tenth of the contracted area. They also used the highly toxic chemical defoliant PCP (sodium penta-chlorophenol), containers of which may have been left in the area to be flooded. The danger from the decomposition of the forest biomass will become apparent only with time. The lake also flooded a small stretch of the Transamazon highway, necessitating relocation of that part.

The Tucuruí dam is at least producing a large quantity of energy from an area that is not a center of endemism or of high biodiversity. On the other hand, the recently completed Balbina dam on the Rio Uatamã north of Manaus is a real environmental disaster. This dam has flooded a larger area of rain forest than Tucuruí and the lake covers much of the territory of the Atroaris-Waimari Indians. The lake is now a sea of dead rain forest trees standing with only the lower part of their trunks in water because the water depth is so shallow. Balbina produces only 8 percent of the energy that Tucuruí generates, yet it has caused a great deal more environmental damage in an area of extremely high biological diversity. This ecological disaster, partially funded by international capital, was allowed to take place because one politician, the governor of the state of Amazonia, had staked his reputation on it.

The serious effects of the Tucuruí and Balbina dams are minimal compared with what could have happened if an extraordinary plan of the Hudson Institute had been implemented.[23] This was a proposal to dam the main Amazon River to produce a vast inland sea! But even without the Hudson Institute's dam, the hydroelectric dams are one of the contributing causes of habitat destruction and extinction in Amazonia.

There are many mining projects scattered around Amazonia. Some have been more destructive than others, but until recently they have destroyed comparatively small areas of forest. The gold mine at Serra Pelada (figure 3.12) in the south of Pará has drawn much attention because a whole hill has been dismantled to remove one of the richest gold deposits ever discovered. The mining project that is likely to cause the most damage because of its size is that of Serra Carajás in Pará, the world's largest iron deposit. The project will use energy from the Tucuruí dam to power its work, and a new railroad has been built from Carajás to the coastal part of the São Luís in Maranhao. The summit of Serra Carajás contains various endemic species of plants confined to the ironstone Canga Formation. These may be condemned to extinction by this operation. The Carajás project, however, is one of the best monitored in terms of environmental controls. The environmental department of the Carajás mining project has been well funded and is a fine example of what can be done. The work of its director, Maria de Lourdes Davis de Freitas, is helping to avoid the

Figure 3.12
The gold mines at Serra Pelada in Pará State of Brazil completely dismantled this hill in one of the biggest gold rushes of this century. Photograph © 1984 by the Cousteau Society, Inc., a nonprofit environmental research organization located at 930 West 21st Street, Norfolk, Virginia 23517. Reprinted by permission.

disaster that might have happened. The government has much broader plans for the development of the Carajás region (Projeto Grande Carajás), and this is likely to cause the vast deforestation and species loss, not the work of the iron mine. The worst effect of the Carajás iron mine has been the establishment of a number of pig iron smelters along the railroad. The smelters are fired by charcoal, and attempts to produce this from sustainable plantations have not been successful so even more rain forest is cut to smelt the ore.

Some other mines have instituted interesting environmental projects. For example, the bauxite mines of the Trombetas River have employed an ecologist, Henry Knowles, to work on the rehabilitation, using native-species of trees, of areas destroyed by their open-cast mining. Tree nurseries are growing many native trees for replanting after the mining is finished. The work of Knowles is one of the best examples of reforestation in Amazonia.[24] It is working well because, from experience, he learned that the thin layer of topsoil from the mine areas must be preserved and returned to the reforested areas. The same bauxite company in the Trombetas region began by polluting the river and severely affecting its fisheries with mine tailings, but extremely good pollution controls have now been installed whereby the tailings are returned to vast artificial lakes that will be reforested once they are full.

There have been several instances of deforestation for large forestry projects in which large tracts of rain forest were cut down to grow a single species. The first one was Henry Ford's Fordlândia near the Tapajós River. This vast rubber plantation was started in 1924 to reestablish rubber as an important crop in its native region. Ford built a whole town, a railroad, and a plantation of over a million trees. The multimillion dollar project failed because the rubber trees were attacked by a fungus and because riverine, flooded forest varieties were planted on the upland terra firme. The leaf rust fungus (*Microcyclus ulei*) is also a native species of Amazonia that has evolved over the years with the native rubber trees. In the forest, however, rubber trees are usually spaced 50–100 meters from their nearest neighbor, lost in a diversity of other tree species. In this natural situation the fungal spores are not often carried from one tree to another, and so the disease is of little consequence. But put many trees next to one another and the spores can be carried from rubber tree to rubber tree, devastating a plantation. For this reason Henry Ford gave up his rubber planation in 1946, after investing $30 million. In tropical Asia, where the fungus does not exist, rubber plantations work well.

The losses of Henry Ford were small compared with those of Daniel K. Ludwig and his Jari forest project. In 1967 Ludwig began a vast plantation in the region of the Jari River, which divides the state of Pará and the territory of Amapá. He planned to plant a "miracle" fast-growing timber tree, *Gmelina arborea,* in his 2.5-million-acre estate. Mistakes were made early in the project, such as clearing with bulldozers and removing topsoil. Land had to be cleared by hand and the gmelina trees planted among the debris. They also planted on sandy soils unsuitable for gmelina and had to abandon large areas because their property runs through the boundary of the lowland Amazon alluvial clay soil and the sandy soil of the Guiana shield. Eventually they planted pine trees on the sandy soils, gmelina on the richest soils, and eucalyptus in intermediate areas. Ludwig's project is probably best known because it floated a $200 million, 17-story paper mill from Japan to Jari, as well as an equally large power plant. The mill is one of the most modern in the world. It is amazing to enter the control room in the middle of the Amazon jungle and see the vast array of computer equipment that allows only six people to control the entire operation. Ludwig's investment was well over a billion dollars, but in 1982 he sold out to a consortium of Brazilian companies for $400 million and the assumption of his debt on the mill. With Ludwig's $600 million loss, the Jari project can hardly be called a success even if it becomes profitable to the present owners. The pulp mill continues to function and to produce much pulp for its new owners, but their profit is from kaolin clay, not from pulp.

These examples are surely enough to show that paradise is being lost at a rapid rate. Fortunately the new decade of the nineties has seen a slowing of the rate of deforestation. Nevertheless, the destruction is a cause for alarm for those who are concerned with the preservation of the genetic resources of the world, the stability of world climate, and the stewardship of the soils of our planet.

These examples are only from the Brazilian Amazon. Although at a less rapid rate, some deforestation is occurring in other Amazonian countries. We could also consider the effects of oil fields in the Peruvian and Ecuadorian Amazon or of the road into the Peruvian Amazon and subsequent installation of cattle pastures and coca plantations there. The forests of Amazonian Bolivia are likewise being cut at an increasing rate to make way for cattle.

Another aspect that I have touched on only briefly is the effect of the deforestation on the Indian peoples of the region. Some of the projects already described have had a largely destructive impact on the Indians, particularly on the tribes whose territory was flooded by the Tucuruí and Balbina hydroelectric dams and the Nhambi-quara Indians, who have the misfortune of living near BR-364, the road across Rondônia that encroaches on some of their sacred sites. The famous Auca (Waorani) Indians of Ecuador were unfortunately situated in a region where oil was discovered. Some tribes are being acculturated, others are going extinct. The loss of these human cultures is not only a moral and ethical tragedy but also a loss of a whole body of information that might have taught us how to create sustainable-use systems in the region.

Species Destruction

So far we have considered mainly large-scale projects that destroy whole habitats and consequently whole cohorts of species. Projects such as the Tucuruí dam, the colonization of Rondônia, and the Jari forestry project devastate entire inhabitants. We could equally well look at the fate of a large number of individual species that have suffered and become threatened and endangered as a result of the modern developments in Amazonia.

A tragic loss is that of the Amazon manatee (*Trichechus inunguis*), or sea cow, one of the most important components of aquatic ecosystems because of the vast quantities of vegetation that they eat. These animals have been over hunted for their delicious meat and oil. They are now severely endangered, and few individuals remain. Their only hope is that they are the focus of a research program of the National Amazon Research Institute (INPA) in Manaus, which is making great efforts to learn about their ecology and physiology so that they can be bred. Experiments have included trying to populate at least one lake behind a hydroelectric dam (Curuá-una) with manatees.

The giant otter (*Pteroneura brasiliensis*) is the largest otter in the world. No one can forget meeting a pack of these curious animals swimming in a river as they pop up beside one's canoe and bark. I will never forget the three times I have been fortunate enough to have had this increasingly rare experience. Their very curiosity makes them a sitting target for the hunter in search of their valuable fur.

Several of the Amazon side-necked turtle species are endangered because of excess exploitation. These creatures do not stand a chance because people prey on both the grown turtles and their eggs laid in the sand of river beaches during the dry season. The giant river turtle (*Podocnemis expansa*), once presented in uncounted numbers, is near extinction.

The various species of crocodilians, large cats such as the jaguar, and large constricting snakes are all sought after for their skins or fur. Although brazil's laws protect these animals, the control of hunting in a region as extensive as Amazonia is impossible, and many pelts are smuggled out through such countries as Bolivia, where there is no protective legislation.

The overexploitation is not confined to animal species. Useful species of trees are subject to such "mining." A good example is rosewood (*Aniba rosaeodora*), which is the source of rosewood oil. Small mobile distilleries are set up in the forest, and cutters are sent out to fell every tree around. Once all rosewood trees have been eliminated locally, the stills move on to other locations to continue the extermination of this species. Yet this kind of destructive exploitation is totally unnecessary. Rosewood can be grown in plantations, as has been demonstrated by INPA. It is also possible to harvest oil from leaves and small branches without destroying the tree. Moreover, there are other sources of the essential oil, linalol, including a common weedy species of secondary forest tree, *Croton cajucara*.

Trees of the species of sorva (*Couma*) are cut down to extract latex, which is used in chewing gum. Again, sorva cutters are penetrating remote areas of otherwise little-disturbed forest to fell systematically all the trees of *Couma*. A visit to one of the processing factories of this latex in Manaus reveals the enormous number of *Couma* trees that are being cut down without being replaced. The massaranduba tree (*Manilkara huberi*), another species that is felled for its latex, is experiencing a similar fate.

Ironically, those species that are most threatened with extinction are also some of the most useful. They are threatened because they are being mined and not managed. In many cases it is possible to farm or cultivate these most useful of Amazon species. It can only be hoped that efforts of the various research institutions, such as INPA in Brazil and IVIC in Venezuela, will help to solve basic management problems in time to prevent the disappearance of these valuable species from the Amazon forests.

Despite all the destruction that I have recounted, there do remain a vast wealth of relatively undisturbed forest and of biological species in Amazonia, and a remnant of the Indian population.

Up to 1992 only 12 percent of Amazon had been totally deforested. There is still, therefore, much forest to save, and many species—including some whose resource value will add to the ways in which we can with prudence use the Amazon region in the future. The challenge for us today is not to lament over the existing destruction but to think of creative ways in which we can regain the paradise.

Satan, in Milton's *Paradise Lost,* felt his efforts were in vain, Milton wrote:

So having said, a while he stood, expecting
Their universal shout and high applause
To fill his ear, when contrary he hears
On all sides, from innumerable tongues

A dismal hiss, the sound
Of public scorn.
(*Book X, lines 504–509*)

Well-meaning industrialists may feel a little like Satan did. It may come as a surprise that conservationists believe that developers tearing down the forest are committing a crime. Projects such as Ludwig's Jari forest have been hailed as great humanitarian enterprises. And Ludwig himself was the recipient of an honorary degree from Johns Hopkins University for his pioneering capitalist enterprise in the Amazon Basin. Fortunately the destruction of rain forest has evoked considerable public outrage over the last decade. The pleas of the conservation community have been partially successful, but there is no room for complacency since the situation could easily reverse. The preservation of the Amazon forest is so vital that we should increase our efforts before it is too late. We are probably the last generation to have the opportunity to conserve the biodiversity of the Amazon forest.

Let us remember as we return to Milton's *Paradise Lost* that Satan was powerful in heaven before the Fall and already had almost everything he could have wanted. The Fall came because he questioned the one unquestionable authority rather than work within an incontrovertible system. Satan's hubris is like humanity's own. The rain forest can provide us with a wealth and richness of life that we have barely begun to tap. The proviso is that, to utilize these riches with any permanency, our articles must absolutely be on nature's terms. There is nothing, it now appears, to stop us from totally despoiling this resource, but, if we do so, we must pay the price. Like Satan, we wield enormous power, but there are some simple facts of life that we cannot change, no matter how much we want to, without destroying that same life and bringing ourselves to ruin in the process. Failure to comprehend and comply will surely bring about our own great fall and with it an unprecedented mass extinction of species—one even greater than when the world lost its dinosaurs. Today we stand at the brink of disaster in Amazonia. Will we follow the path of numerous self-interests and short-term gains into the abyss, or will common sense prevail and allow us to work out a plan for accepting and using Amazonia on nature's terms?

The continued plundering of Amazonia will begin to affect seriously the physical environment that sustains the forest. It will not change the world's oxygen balance, as has so often been misstated; instead it will produce other equally grave effects. Cutting down the Amazon forest will add to the atmospheric carbon dioxide and the possibility of a warming world climate through the so-called greenhouse effect. Knowing that 50 percent of the rainfall in Amazonia comes from the transpiration of the trees themselves points to another consequence of deforestation (see note 4). If you replace the forest with grassland with a much reduced leaf area, then the rainfall will be greatly reduced. With such a reduction, the amount of savanna and other arid vegetation types will increase.

The skeptic in Brazil will ask, with considerable justification, what about the mess we have made here? The United States is prosperous partly because of its original deforestation and our destruction of other native ecosystems and resources. It *is* different in the tropics, however, with the poor soils, high species diversity, and fragile ecosystem. Temperate ecosystems, such as the eastern deciduous forest of North America, can often recover from serious damage. Tropical rain forest cannot. Most important, we now have available much more scientific knowledge than we did at the time the North American forests were felled. It is not necessary to cut down all the Amazon forest to sustain the population of Brazil. We are now beginning to learn how we might achieve a balance between conversation and utilization before it is too late.

In writing this account of a lost paradise, I do not mean to criticize what is already happening in a positive way in Amazonia. Fortunately, in several Amazon countries, significant things are occurring both in the conservation of natural areas and the creation of sustainable systems of land use. Peru, for example, has set up the large Manu National Park in its Amazon region, Venezuela, the vast Canaima National Park, and Brazil, the Tapajós National Park as well as several ecological reserves. Colombia has given a third of her Amazon territory back to the Indians, who now hold the title to the land. Brazil has now removed the fiscal incentives for cattle pasture in the Amazon region and, following the assassination of the leader of the rubber tappers union, Chico Mendes, has created a series of extractive reserves. These reserves, where the tappers can exploit forest products such as rubber latex and Brazil nuts but not deforest, have themselves helped to halt deforestation in some areas, particularly in the states of Acre and Amapá. This is an excellent start for the conservation of Amazonia,[25] but the present system of parks does not cover nearly enough area to avoid species extinction on a large scale (figure 3.13). Furthermore, many declared parks and reserves exist only on paper and confer no actual protection on the site. We need to assist the Amazon countries to build on the firm foundation they have begun, to produce a network of conserved areas that protects enough area of forest and as varied a range of vegetation types and local habitats as is possible.

Figure 3.13
This spot, only 120 kilometers north of Manaus at Igarapé dos Lajes, is formed by a rock outcrop and a black-water stream. It is a site of much plant endemism and could have made a beautiful park. Instead it has been decimated by settlers. Photograph by Ghillean Prance.

Similarly, there are at least some efforts already under way in most Amazon countries to study and produce sustainable yield systems of agroforestry and other continuous systems, such as the seasonal use of nutrient-rich floodplains for agriculture. This knowledge is coming about through both ecological studies of indigenous peoples[26,27] and well-planned agroforestry programs, especially in Peru and Brazil. Some crops have been relatively successful in Amazonia, including jute (*Corchorus capsularis*) and Malva (*Urena lobata*), grown for fiber on the floodplains of Brazil, and black pepper, grown in conjunction with chicken farming, which produces manure fertilizer. Although these examples were successfully introduced by Japanese farmers, such crops are not without their problems. The task of retting the jute in water is an unhealthy, hazardous occupation, as is work in the jute factories, where there is no adequate protection from the dust. In the case of black pepper, crop disease caused by fusarium has rendered many areas unproductive. The indigenous experience and modern research is showing that polyculture agroforestry systems are most likely to succeed, rather than any type of monoculture, whether it be gmelina trees or the pasture grass *Brachyaria*.

Paradise Regained: The Amazon of the Future

I who e're while the happy Garden sung,
By one mans disobedience lost, now sing
recover'd Paradise to all mankind.
(*John Milton, Paradise Regained, Book I, lines 1–3*)

Balance of Conservation and Utilization

Do we have to let the destruction continue, or can we regain a little of
this paradise? I have shown that many but not all of the projects that are
causing deforestation are for replacement systems that we can with justification call ecologically unsound: the sustenance, for example, of one cow on
five acres instead of over 1200 individuals of 200 species of trees. The
Amazon of the future will not be a paradise for its human population if it
cannot sustain it. The answer to the present destruction, therefore, is not
to create a vast biological reserve as a playground for naturalists and rich
tourists from more developed countries. Paradise can be regained only
through a balance of conservation and utilization. Too often these two
words are at opposite extremes. On a worldwide basis this is beginning to
change, especially since the production of the World Conservation Strategy
by the International Union for the Conservation of Nature and Natural
Resources (IUCN), the World Wildlife Fund (WWF), and the United
Nations Environment Program (UNEP) in collaboration with the Food
and Agricultural Organization of the United Nations (FAO), its 1991
revision entitled "Caring for the Earth," and the report of the Brundtland
Commission "Our Common Future." These international reports and the
1992 United Nations Conference on Environment and Development
(UNCED) have progressively put more emphasis on sustainable land use
together with the conservation of biodiversity.[28]

The achievement of a balance between conservation and sustainable
use is certainly the most logical step for the future of Amazonia.[29] There
we need to remember the large size of the indigenous population that the
region sustained before it was conquered by Westerners. We therefore need
to learn as much as we can from what is left of these indigenous cultures;
this means that ethnobotanical and especially ethnoecological studies of the
Amazonian Indians are an urgent priority.

It is also urgent to inventory and discover the useful plants and animals
of the Amazon and of tropical rain forests of other parts of the world. It is
among these unknown and underexploited species that we will find ones
more suitable for the region than cattle or the gmelina tree. It is no
coincidence that the main emphases of the newly formed Institute of

Economic Botany of the New York Botanical Garden are the study of indigenous agricultural systems and the search for new food and fuel plants from the tropical forest. That is why we have investigators studying the fruit trees of the Peruvian Amazon, the agroforestry systems of the Amuesha nation in Peru, the ethnobotany of the Chácobo Indians in Bolivia, the economic plants of the Ecuadorean Amazon, and the economics and botany of the babassu oil palm (figure 3.14) in Brazil.[30] But this is only a beginning. There is a desperate need for a basic inventory of all the plants of Amazonia, not just those of known economic potential.

Experience has shown that in most of Amazonia it is difficult to make permanent cattle pasture. We must develop sustained yield systems that continue to support people through the years. The future should be much more with trees and perennial crops than with open clearing and exposure of the fragile soil to grassland or annual crops.

Why use the cow, an animal that does not belong in the rain forest habitat? Unlike the drier African grasslands, which are suitable for large

Figure 3.14
The babassu palm (*Orbignya phalerata*) forms huge natural stands in the southeastern portion of Amazonia. It sustains over two million people who work on the extraction of oil and other babassu products, such as charcoal and meal. Photograph by Ghillean Prance.

browsing animals, the Amazon does not have a heritage of many such animals. There are no herds of elephants, giraffes, wildebeests, zebras, or cape buffalo roaming the Amazon region. Instead there are only tapir and a few deer. Nature is telling us something. Large browsers have not evolved in Amazonia, and we should follow nature's hint. Use the Amazon primarily to produce trees, not beef. Why not experiment with limited development of other meat-producing animals more suitable for the region? Water buffalo have already been successful in the Amazon delta region, especially on the Switzerland-sized Marajó island at the mouth of the Amazon. Venezuela is experimenting with the commercial production of capybara (*Hydrochoerus hydrochaeris*; figure 3.15),[31] the world's largest rodent and an Amazonian native. Capybara meat is highly esteemed. Turtle farming is underway in Belém. Brazil could produce turtle meat without depleting the wild stock of these severely endangered animals. We should also experiment with the local species of deer, tapir, agouti, and other native animals that might be farmed in a less destructive way than cattle.

Figure 3.15
The capybara (*Hydrochoerus hydrochaeris*), the world's largest rodent, is common in Amazonia. Its good meat has caused it to be overhunted but has also led to movements in Brazil and Venezuela to domesticate this animal as a substitute for cattle. Photograph by Loren McIntyre.

The lesson from Rondônia, mentioned as an example of destruction, is that some uses of the rain forest area are working. Conservationists need to examine some of the successful agricultural projects of Rondônia and see the number of people who have been successfully settled there before they condemn every deforestation project out of hand. In a visit to Rondônia, in late 1984, I was impressed by some of the cacao plantations, rice cultivation projects, and even a few of the cattle ranches among many others that were obvious failures. Those of us who are concerned about deforestation and the conservation of the genetic resources of Amazonia are quick to mention the failures, for example, Fordlândia and the Transamazon colonization, but slow to acknowledge that some projects that have replaced primary forest are in fact working and providing a good living for their owners. These projects are in the minority, but they do demonstrate that some people have mastered techniques to use the fragile Amazon soil. One of the tragedies of rain forest conservation is the gap between the conservationist and the agronomist or forester. We need at least to examine what they have to say to begin planning with them. We must keep informed about which development projects are working as well as about the ones that are not.

As with examples of destruction, there is a long list of constructive programs that should be called to attention; again only a few examples can be cited to indicate the direction in which future development of the Amazon region is more likely to succeed.

The residents of the small town of Tamshiyacu in the Peruvian Amazon, mostly of Indian extraction, have developed an agroforestry system based on a wide diversity of crops and products. An approximately thirty-year cutting cycle of trees is followed.[26,32,33] As a result, these people are extremely well off in comparison to most rural Amazonians. Their largest crop, the umari (*Poraqueiba*), is a fruit tree that yields throughout most of the cycle but is finally sacrificed as a source of firewood. The Brazil nut trees, which take much longer to grow and yield, are one of the other trees interspersed into the system. They are successfully pollinated in such a mixed forestry system, and they also obtain sufficient nutrients from the soil in this managed forest. Eventually they yield a high quality timber as well. One of the keys to the Tamshiyacu success is a diversity of crops rather than a plantation of one species, something that is much more similar to the natural Amazon ecosystem.

The original indigenous populations on the Amazon were largely riverine floodplain-based cultures, such as the Omagua and Tapajós Indians described by Meggers (see note 2). They cultivated between the floods, had elaborate storage techniques for preserving food through the flood season, and used the fishery resources of the river. The riverside soils along white-water rivers are different from those of the terra firme because of the effect of annual flooding, with its deposition of nutrients from the alluvial matter. Table 3.2 shows a comparison of soils in two areas of floodplains and terra firme. The riverside soil is much more appropriate for continued agriculture and is even less acidic.

Recently, various agricultural projects have been established on flood-plain areas, following a suggestion made years ago by one of the sages of Amazon development, Felisberto Camargo.[34] The city of Manaus used to have problems with the production of lettuce, chives, and other green vegetables. A continuous production now comes from the floodplain near Iranduba and other similar areas. This is in marked contrast to the complete failure of a city government project on terra firme at Iranduba. The Iran-duba projects are a good example of the contrast between production from the floodplains, or várzea, and that of terra firme.

There are problems with floodplain agriculture, however, and one is the damage to fisheries. Recent work, especially by Goulding,[35] has revealed the large number of Amazonian fishes that eat fruit and leaves and that mainly take their nourishment in the flood season by swimming through the flooded forest (figure 3.16). Because the fishery is one of the other priceless resources of the Amazon, floodplain agriculture must not be allowed to interfere with the life cycle of some of the most appetizing of all Amazon fishes, by clearing the forest on which they depend. Conse-quently, when advocating the use of the floodplain, one must also ensure preservation of adequate areas of forest for the fishes or even the interspers-ing of agricultural areas with tree crops that bear fruit eaten by the fishes. An analysis of the fruits that the fishes eat shows that many are from trees of considerable utility, such as the rubber tree (*Hevea*), the andiroba (*Carapa guianensis*), a source of timber and medicinal oil, and the jauari palm (*Astrocaryum jauari*), a source of thatching material and fiber. Fields in which lettuce is grown interspersed with rubber and andiroba trees will be far less damaging to the fisheries than open fields without trees.

I have already drawn attention to the reforestation project of Henry Knowles at a bauxite mine on the Trombetas River. One of the most important ways to preserve the Amazon forest is to reforest the areas that have already been destroyed in such a way that they become productive for timber, agroforestry, and other appropriate native crops such as guaraná and cupuaçú (*Theobroma grandiflorum*). Pioneer work in reforestation has been carried out by ecologist Christopher Uhl and his associates who are showing the way in which this can be done.[36]

Table 3.2.
Comparison of floodplain and terra firme soils near Manaus Brazil[a]

Sample	Habitat	pH	Phosphorus (ppm[b])	Potassium (ppm)	Calcium (wt. %)	Magnesium (wt. %)	Aluminum (wt. %)
	Floodplain						
401	Ariau–Várzea floodplain	5.3	70	45	7.0	2.4	0.3
402	Ariau–Várzea floodplain	5.3	85	62	8.5	2.3	0.3
409	Curua–Una floodplain forest	6.6	65	46	4.5	0.8	0.0
	Cattle pasture on terra firme						
403	T. Loureiro cattle pasture	4.0	2	12	0.3	0.1	2.0
404	T. Loureiro cattle pasture	4.0	3	20	0.5	0.2	2.0
	White sand						
405	Inpa white sand campina	4.7	2	8	0.4	0.2	0.7
	Terra firme						
407	Inpa terra firme forest (KM45)	3.5	1	8	0.3	0.1	2.8
408	Inpa terra firme forest (KM45)	3.5	1	12	0.3	0.1	3.0

Data from F. Magnani, based on analysis made at INPA, 18 November 1982.
a. For comparison the average agricultural requirements for Amazonian soils are as follows:
Potassium: low, 10; high, 30.
Phosphorus: low, 46.8; high, 117.
Calcium: low, 2.0; high, 5.0.
Magnesium: low, 0.5; high, 1.0.
Aluminum: Above 0.5 is often toxic to plants.
b. ppm = parts per million.

Figure 3.16
A view of the forest in the floodplain, where for part of the year fish graze among
the trees. Photograph by Ghillean Prance.

The challenge of the future is to design and ensure implementation
of these constructive systems rather than the more destructive ones more
often selected. This should involve much more forestry and much less
agriculture than is currently being practiced. If some of the region can be
rendered productive on a long-term basis to sustain the nourishment and
livelihood of the Amazon region, then other large areas can be conserved
in their pristine state for the preservation of the species pool of the region,
the maintenance of the fragile soils, and the continuation of the high annual
rainfall that is so largely dependent on the transpiration of the forest.

Irrevocable loss *can* be averted in the Amazon region, but every year
we get nearer to the brink of losing what will soon be the world's last rain
forest paradise. We must use every influence we have to help the Amazon
forest countries in a constructive way to use but not abuse their greatest
resource, the Amazon.

Many are the ancient temples and cathedrals of which only the ruins
remain. There we can pace between the sontes that once supported soaring
buildings. We can sadly wonder and postulate what these places were once
like, yesterday's glory and today's "bare ruined choirs." They fell victim to
new ways, new faiths, new conquerors.

Successive waves of new ways, new faiths, and new conquerors in various guises threaten the Amazon rain forest. That long-lasting ever-evolving paradise is in danger. Shall it be said of us, like Adam and Eve leaving the garden,

Some natural tears they drop'd, but wip'd them soon

as

They hand in hand with wandering step and slow,
Through *Eden* took thir solitaire way?
(*John Milton,* Paradise Lost, *Book XII, lines 645, 648–649*)

Or shall we courageously employ our knowledge, skill, creativity, and time to retain Nature's great architecture and to regain for all time the Amazon paradise?

Notes

1. N. Myers, *The Sinking Ark: A New Look at the Problem of Disappearing Species* (Oxford and New York: Pergamon Press, 1979).

2. B. J. Meggers, *Amazonia: Man and Culture in a Counterfeit Paradise* (Chicago and New York: Aldine Atherton, 1971).

3. E. Salati, T. E. Lovejoy, and P. B. Vose, "Precipitation and water recycling in tropical rain forests with special reference to the Amazon Basin," *Environmentalist* 3(1) (1983), 67–72.

4. E. Salati and P. B. Vose, "Amazon Basin: A system in equilibrium," *Science* 225 (1984), 129–138.

5. H. W. Bates, *The Naturalist on the River Amazons* (London: J. M. Dent & Sons, Aldine Press, 1969), 406–407.

6. P. Fleming, Introduction to the 1969 edition of H. W. Bates, *The Naturalist on the River Amazons* (New York: Dover).

7. G. T. Prance, W. A. Rodrigues, and M. F. da Silva, "Inventário florestal de uma hectare de mata de terra firma km 30 Estrada Manaus—Itacoatiara," *Acta Amazonica* 6 (1976), 9–35. A. H. Gentry, "Tree species richness of upper Amazonian forests." *Proceedings of the U.S. National Academy of Sciences* 85 (1988), 156–159.

8. A. H. Gentry, "Patterns of neotropical plant species diversity," *Evolutionary Biology* 15 (1982), 1–84.

9. A. A. Federov, "The structure of the tropical rain forest and speciation in the humid tropics," *Journal of Ecology* 54 (1966), 1–11.

10. J. Haffer, "Speciation in Amazonian forest birds," *Science* 165 (1969), 131–137.

11. G. T. Prance, ed., *Biological Model of Diversification in the Tropics* (New York: Columbia University Press, 1982).

12. G. T. Prance, "Forest refuges: Evidence from woody angiosperms," in *Biological Model of Diversification in the Tropics,* 137–158.

13. J. H. Connell, "Diversity in tropical rain forests and coral reefs," *Science* 199 (1978), 1302–1310.

14. D. H. Janzen, "Herbivores and the number of tree species in tropical forests," *American Naturalist* 104 (1970), 501–528.

15. P. S. Ashton, "Speciation among tropical forest trees: Some deductions in the light of recent evidence," *Biological Journal of the Linnean Society* 1 (1969), 155–196.

16. P. W. Richards, "Speciation in the tropical rain forest and the concept of the niche," *Biological Journal of the Linnean Society* 1 (1969), 149–153.

17. G. T. Prance, "Notes on the vegetation of Amazonia III. The terminology of Amazon forest types subject to inundation," *Brittonia* 31 (1979), 26–38.

18. G. T. Prance, "The pollination and androphore structure of some Amazon Lecythidaceae," *Biotropica* 8 (1976), 235–241; S. Mori, G. T. Prance, and A. B. Bolten, "Additional notes on the floral biology of neotropical Lecythidaceae," *Brittonia* 30 (1978), 113–130. B. W. Nelson, M. L. Absy, E. M. Barbosa, and G. T. Prance. "Observations on flower visitors to *Bertholletia excelsa* H.B.K. and *Couratari tenuicarpa* A.C. Smith (Lecythidaceae). *Acta Amazonica* 15 (1/2 Supplemento (1986), 225–234.

19. D. A. Posey, "The keepers of the forest," *Garden* 6 (1982), 18–24."

20. D. A. Posey, "A preliminary report on diversified management of tropical forest by the Kayapo Indians of the Brazilian Amazon," *Advances in Economic Botany* 1 (1984), 112–126.

21. B. M. Boom, "'Advocacy botany for the neotropics,'" *Garden* 9(3) (1985), 24–28, 32. B. M. Boom, "Ethnobotany of the Chácobo Indians." *Advances in Economic Botany* 5 (1987), 1–68.

22. P. M. Fearnside and E. Salati "Sem florestas na proxima decada? Rondônia," *Ciencia Hoje* 4(19) (1985), 92–94.

23. R. Panero "A South American "Great Lakes" system," *Hudson Institute Report* HI 788/3-RR (1967).

24. G. T. Prance, "Give the multinationals a break" *New Scientists* 1683 (1989), 62.

25. G. B. Wetterberg, G. T. Prance, and T. E. Lovejoy, "Conservation progress In Amazonia: A structural review," *Parks* 6(2) (1981), 5–10.

26. W. M. Denevan, J. M. Treacy, J. B. Alcorn, C. Padoch, J. Denslow, and S. F Paitan. "Indigenous agroforestry in the Peruvian Amazon: Bora Indian management of swidden fallows," *Interciencia* 9 (1984), 346–357.

27. W. E. Kerr and D. A. Posey, "Informações adicionais sobre a agricultura dos Kayapó," *Interciencia* 9 (1984), 392–400.

28. IUCN, UNEP, and WWF, World Conservation Strategy. 1980. IUCN, UNEP, WWF. "Caring for the Earth: a strategy for sustainable living," Gland (1991). World

Commission on Environment and Development. "Our common future," Oxford University Press (1987).

29. G. T. Prance, "Future of the Amazonian rainforest." *Futures* 22 (1990), 891–903.

30. G. T. Prance and M. J. Balick, eds., "New directions in the study of plants and people," *Advances in Economic Botany* 8 (1990), 1–278. K. H. Redford and C. Padoch, eds., "Conservation of Neotropical forests: Working from traditional resource use." (New York: Columbia University Press, 1992).

31. J. Ojasti, *Estudio biológico del chiguire o capibara.* Report of Fondo Nacional de Investigaciones Agropecuarias, Caracas, Venezuela (1973).

32. S. F. Paitan, "Analysis of Old Bora swidden fallows," *Advances in Economic Botany* (to be published).

33. C. Padoch and W. de Jong, "Production and profit in agroforestry practices of native and ribereño farmers in lowland Peruvian Amazon," pp 102–114. In J. O. Browder, ed., *Fragile lands of Latin America* (Boulder: Westview Press 1989.)

34. F. C. Camargo, "Reclamation of the Amazon Flood-lands Near Belém," in *Proceedings of the U.N. Scientific Conference on the Conservation and Utilization of Resources,* Lake Success, New York, August 17-September 6, 1949 (New York: United Nations, 1951), vol. 6, 598–602.

35. M. Goulding, *The Fishes and the Forest* (Berkeley, California: University of California Press, 1980).

36. C. Uhl, D. Nepstad, R. Buschbacher, K. Clark, B. Kauffman, and S. Subler, "Studies of ecosystem response to natural and anthropogenic disturbances provide guidelines for designing sustainable land-use systems in Amazonia," pp 24–42, in: A. B. Anderson, ed., *Alternatives to Deforestation* (New York: Columbia University Press, 1990). D. Nepstad, C. Uhl, and A. Serrão, "Surmounting barriers to forest regeneration in abandoned highly degraded pastures: A case study from Paragominas, Pará, Brazil," pp 215–229, in A. B. Anderson, ed., *loc. cit,* 1990.

Suggested Readings

A. B. Anderson, ed., *Alternatives to deforestation: Steps, for sustainable development.* New York: Columbia University Press, 1990.

W. Bates, *The Naturalist on the River Amazons,* reprinted 1975. New York: Dover Publications.

C. Caufield, *In the Rainforest: Report from a Strange, Beautiful, Imperiled World.* New York: Knopf, 1985.

S. H. Davis, *Victims of the Miracle: Development and the Indians of Brazil.* Cambridge: Cambridge University Press, 1977.

J. S. Denslow and C. Padoch, eds., *People of the tropical rainforest.* Berkeley: University of California Press, 1988.

R. J. A. Goodland and H. S. Irwin, *Amazon Jungle: Green Hell to Red Desert?* Amsterdam and New York: Elsevier, 1975.

M. Goulding, *The Fishes and the Forest.* Berkeley, California: University of California Press, 1980.

S. Hecht and A. Cockburn, *The fate of the forest: Developers, destroyers and defenders of the Amazon.* London: Penguin, 1990.

J. Kandell, *Passage Through El Dorado. Traveling the World's Last Great Wilderness.* New York: Morrow, 1984.

B. J. Meggers and C. Evans, *Amazonia: Man and Culture in a Counterfeit Paradise.* Chicago and New York: Aldine & Atherton, 1971.

E. F. Moran, *Developing Amazonia.* Bloomington, Indiana: Indiana University Press, 1981.

N. Myers, *The Sinking Ark.* Oxford and New York: Pergamon, 1979.

N. Myers, *The Primary Source: Tropical Forests and Our Future.* New York and London: W. W. Norton, 1984.

G. T. Prance and S. Cunningham, *Out of the Amazon.* London: HMSO, 1992.

G. T. Prance and T. E. Lovejoy, eds., *Key Environments: Amazonia.* Oxford and New York: Pergamon, 1985.

K. S. Redford and C. Padoch, eds., *Conservation of Neotropical forests: Working from traditional resource use.* New York: Columbia University Press, 1990.

R. D. Stone, *Dreams of Amazonia.* New York: Elisabeth Sifton Books, Viking, 1985.

4

Vanishing Species in Our Own Backyard: Extinct Fish and Wildlife of the United States and Canada

James D. Williams and Ronald M. Nowak

Lost in the heated debate of how to save our tropical forests is the severity of ongoing, human-caused extinction in our own backyard: the temperate and boreal regions of North America. Compared with the tropical world, the temperate forest hosts a fewer number of species, though each claims a far greater number of individuals spread over a wider area. With this kind of profile, we might expect temperate wildlife to be much more resistant to the pressures of extinction. A calamity in one part of its range might leave enough individuals to form a breeding population and begin building anew. But despite its plausibility, such simple logic does not hold, as you can see from the following catalog of events. (See also table 4.1 at the end of this chapter for a complete list of extinct vertebrates since 1492.)

Ways Things Can Go Extinct

Direct Take or Killing

Like extinction, the death of an individual is a natural process. Few of us would argue for premeditated murder, however. The rules of our society generally encourage as long a life as possible for its members, and, unless there are compelling reasons to do otherwise, the premature extinction of a species is something we try to avoid. Yet deliberate killing by human hunters was for many years the leading known cause of extinction. Humans killed other species to get food, clothing, and other materials. They also killed for recreation and to protect themselves, their property, and their domestic animals.

The extermination of animals by human hunters probably went on for thousands of years in North America. The spread of late Stone Age hunters across the continent at the end of the Pleistocene, for example, is a likely cause of the disappearance of the mammoth, giant ground sloth, horse, and about thirty other kinds of large mammals. The West Indies was the last part of the New World that humans reached. One kind of ground sloth, weighing up to 100 pounds, was said to have survived on Puerto Rico until around the time Columbus arrived in 1493. A giant shrew and four comparatively large rodents, also restricted to Puerto Rico, became extinct at about the same time. Native hunters had probably already reduced their numbers considerably, though their final disappearance could have been caused when the Spanish introduced predatory rats and cats.

Marine mammals were largely unaffected by the wave of extinction that hit the large land species at the end of the Pleistocene. As people became more adept at seafaring and as European explorers, traders, and settlers spread over the globe in the sixteenth and seventeenth centuries, sea-dwelling species came under increasing pressure. They fed hungry sailors and yielded valuable commercial products, such as meat, oil, ivory, and baleen. Each of the three major orders of marine mammals—Cetacea, Pinnipedia, and Sirenia—has one recently extinct North American representative, and each of these was exterminated primarily by the activities of people.

In 1493, Columbus discovered the Caribbean monk seal, the only tropical pinniped in the New World. Hunted relentlessly by the Spanish, it was already rare in the eighteenth century, and it continued to decline subsequently, partly because of the persecution by fishermen, who considered it a competitor. The last recorded Caribbean monk seal in U.S. waters is one killed near Key West in 1922. Although there are still occasional, unverified reports of pinnipeds from remote islands in the West Indies, several recent organized surveys have failed to locate the monk seal. It seems likely that the species became extinct by about 1960. A related species survives in Hawaii, but antagonism from fishermen and other human disturbances make it an endangered species.

At the opposite end of the continent was another recently extinct species, the Steller's sea cow, the largest sirenian known from historical time. In the late Pleistocene this same species apparently inhabited the entire rim of the north Pacific, from Japan to California. Native hunters subsequently restricted its range to the waters off Russia's Commander Islands and probably farther east in the Aleutians. A Russian expedition led by Captain Vitus Bering eventually discovered the Steller's sea cow in the Aleutians when their boat became stranded there in 1741. At that time the total sea cow population was probably not over 1000–2000 individuals. Bering's crew and subsequent visitors to the area slaughtered the relatively helpless animals for meat and hides, and, although there were later reports from various areas, the species was probably extinct by 1768.

The large cetaceans were generally more difficult to hunt, though some coastal-dwelling species were the subject of intense human pursuit from early on. The gray whale, now found only in the Pacific, was formerly represented by an Atlantic subspecies. By A.D. 500 the first of the Atlantic gray whale populations had been wiped out along the coast of Europe, followed some time later by its North American counterpart, an early casualty of the whaling industry that developed here after the settlement of New England. Although records are not precise, the North American gray whale seems to have vanished in the early 1700s, even before the time of lengthy oceanic whaling voyages. Although the Pacific gray whale almost suffered the same fate by 1900, federal protection came just in time, and the population there has increased to almost its original number. These two gray whale subspecies represent extremes in human impact on the large whales. All the other kinds—sperm, sei, fin, blue, humpback, right, and bowhead—declined dramatically because of hunting, especially during the nineteenth and twentieth centuries, but none of them has either completely disappeared (like the Atlantic gray whale) or made a substantial comeback (like the Pacific gray whale).

The demise of several marine birds parallels that of the mammals. Not content with destroying the Steller's sea cow population on Bering Island, hunters and trappers pursued the spectacled cormorant as well. It has not been seen since 1852. The great auk (figure 4.1), a large flightless species resembling a penguin, once inhabited the entire rim of the North Atlantic, just as the gray whale did. Popular for its meat and oil, the auk disappeared from the coast of New England by about 1700. It survived awhile longer farther to the north, but the last member of the species was apparently killed in Iceland in 1844.

By the latter half of the nineteenth century, the emphasis of human hunting had shifted inland. Wild birds became a major commercial source of food. One species, the Labrador duck of the Northeast, was killed off by 1875. By the time of the Revolutionary War, hunters had decimated the heath hen (figure 4.2), also found in the same region. Only a century later, this relative of the prairie chicken had disappeared entirely from the mainland. Although a protected colony seemed to thrive for a while longer on the island of Martha's Vineyard, Massachusetts, the last bird there died in 1932. The fate of the heath hen shows how a species is not out of danger simply because one part of its population is protected and has reached the carrying capacity of its habitat.

Even a vast number of animals over a large geographical area cannot be considered safe. This principle is vividly illustrated by the story of the

Figure 4.1
The great auk, which nested on islands of the North Atlantic and migrated as far
south as Florida in the winter, was hunted to extinction by the mid-1800s. Watercolor by John James Audubon, courtesy of the New York Historical Society, New
York.

passenger pigeon. In the 1800s, passenger pigeons still flocked by the billions
throughout the eastern deciduous forests, but by 1900 they had disappeared
from the wild completely as a result of uncontrolled commercial hunting.
Even though some flocks were still in existence after systematic hunting
had ceased, the species continued to decline, perhaps because its population
was so small that it could not sustain the number necessary for reproduction.
One lone captive bird lived in the Cincinnati Zoo until 1914.

In the late nineteenth century, people considered the feathers and
plumes and sometimes the entire skins of birds fashionable apparel. The
Carolina and Louisiana parakeets (figure 4.3) of the southern United States
were avidly hunted for fashion, for pets, and also for the protection of fruit
crops. Both subspecies vanished from the wild shortly after 1900, and the
last captive bird died in 1918. Plumage hunting caused a drastic reduction
of several other bird species, but the creation of refuges and passage of
protective legislation in the early twentieth century prevented additional
extinctions.

During this same period, a large number of mammals were also being
killed for their meat and skins. Subsistence, recreational, and commercial
hunting all played a part, and by about 1900 most species of big game

Figure 4.2
Audubon noted the rapidly shrinking distribution of the heath hen as early as 1830.
Efforts to save the species during the early part of this century were unsuccessful.
Watercolor by John James Audubon, courtesy of the New York Historical Society,
New York.

mammals and many kinds of furbearers would have been considered endangered or threatened by today's standards. Fortunately, perhaps, most of
the species of value also had extensive ranges and were not restricted to
limited habitats; and this might have helped them survive until the large-
scale conservation movements of the twentieth century. There were some
exceptions. The eastern elk, which once inhabited land from the Great
Plains to the Appalachians, was hunted to extinction by the 1880s, and
Merriam's elk of the Southwest followed shortly thereafter. The tule elk,
found only in the Central Valley of California, was apparently once reduced
to a single pair of animals, but timely protection allowed the eventual
reestablishment of several herds. There is still some doubt, however, about
the future of this subspecies and certain other populations of elk along the
West Coast.

In general, elk, deer, moose, and pronghorn antelope have responded
favorably to human conservation efforts since 1900. Their numbers have
grown substantially since then, and they are playing a natural, ecological
role over vast areas. The plains bison no longer has such a role, but it at

Figure 4.3
A large number of Carolina parakeets were killed for their brilliant plumage, which
was used to decorate ladies' hats during the late 1800s. Watercolor by John James
Audubon, courtesy of the New York Historical Society, New York.

least appears to have been saved from total extinction. Despite their general success, the bison, deer, and pronghorn are each represented by one or more endangered subspecies.

Two other big game species, the caribou and bighorn sheep, were even more susceptible to human disturbance and less responsive to the efforts of conservation. They have not experienced general recoveries since the declines of the late nineteenth and early twentieth centuries, and each is represented by an extinct subspecies. Because the badlands bighorn had a more restricted range and a more accessible habitat than other kinds of bighorn, hunters wiped them out before conservation measures could take effect. The caribou subspecies restricted to the Queen Charlotte Islands off British Columbia was probably unable to adjust to increasing human presence within its limited habitat, and it disappeared by 1935.

The reduction in big game animal numbers and distribution presented their natural predators with a crisis. The same problem could have existed thousands of years before when Stone Age hunters killed off the mammoths and ground sloths, thereby leading to the demise of the dire wolf and saber-toothed tiger. Unlike the situation then, the wolves, cougars, and bears of modern times had an alternative prey available—domestic livestock. Although the degree of predation by wild carnivores has often been exaggerated, it is still considered a serious problem by many ranchers, farmers, and shepherds. The human counterattack was devastating and up to now has been responsible for the extinction of more kinds of mammals in the United States and Canada than any other single factor.

Large mammalian predators generally declined as the frontier advanced to the west and north. Wolves, perhaps because of their dependence on pack structure and their tendency to occupy open habitat, seemed especially vulnerable. The two eastern subspecies of cougar have managed to survive to the present, mainly in Florida and perhaps in the Appalachians, though each probably has only a few dozen individuals. The cougar still occupies most of its original range in the western mountains. The grizzly bear, however, was rapidly extirpated from about 99 percent of its range in the contiguous United States, mainly during the late nineteenth and early twentieth centuries. Its subspecies in California, having been intensively hunted ever since the establishment of cattle ranches there by the Spanish in the late 1700s, seems to have disappeared by 1925.

Complaints by stockmen about wolves, cougars, bears, and other carnivores that hunted on government lands in the West resulted in a large-scale federal predator control program in 1914. This effort was first directed mainly against the gray wolf. By 1940, this species had been practically exterminated in the western contiguous United States, six sub-species having become extinct. Attention then shifted to the smaller red wolf, which still held out in the south-central United States. As its numbers fell, coyotes moved in from the west to fill the vacant predatory niche. The two canines, being closely related and evidently identical in chromosomal structure, began to interbreed, and the remnant red wolf populations gradually faded away. The last pure group of the Texas subspecies, found along the Gulf Coast south of Houston, seems to have been genetically swamped by 1970. The central subspecies (*Canis rufus gregoryi*) survived for a few more years in extreme southeastern Texas and southern Louisiana but probably has disappeared from the wild in the pure form by now. A few dozen individuals were brought into captivity before the complete spread of the hybridization process, however, and only these captive individuals and their progeny have kept the subspecies *gregoryi* off the extinct list.

Habitat Disruption

At present the most serious and frequent threat to the survival of species is the disruption of their habitats. Habitat changes are physical, chemical, and biological, but in most ecosystems the components are so tightly inter-twined that alteration in one results in disruption of others. Physical altera-tions are usually the most obvious: they include clearing of natural vegetation and cultivation of the land, commercial and residential develop-ment, draining and filling wetlands, stream channelization, dams, and struc-tures for water diversion (figures 4.4–4.11). The effects of chemical pollution are often less obvious than physical changes but equally damaging to the ecosystem. Chemical changes resulting from acid rain, improperly treated organic waste, pesticides, high concentrations of heavy metals, toxic wastes, and other industrial and agricultural chemicals can completely de-stroy the biological components of an ecosystem. Biological disruption of an ecosystem often occurs when an exotic or nonnative species is intro-duced. Such introduced species prey on native animals, compete with them for food and space, and interbreed with them.

In North America the problem of habitat disruption began with the settlement of the East Coast during the 1500s and 1600s. The problem did not catch the attention of naturalists until the mid-eighteenth and nine-teenth centuries. During this period, activities such as timber harvest, cultivation of the land, water pollution from slaughterhouses, tanneries, and saw mills, construction of mill dams, and draining of wetlands began to take their toll on the native plant and animal communities. In the aquatic environment, such disruption resulted in the decreased harvest of shad,

Figure 4.4
Comanche Springs and other large springs in west Texas have been pumped dry, totally destroying their aquatic fauna. Photograph by Jim Williams.

Atlantic salmon, and other anadromous fishes. As the human population expanded and moved westward, the loss of natural habitat progressed across the plains through the western mountains to the Pacific Ocean. Not only has habitat disruption persisted to the present day, but it has continued at an increased rate, consuming larger portions of relatively undisturbed habitat and leaving in its wake scores of extinct plants and animals.

The first vertebrate extinction in the United States that resulted from human encroachment on a habitat was probably the giant deer mouse, known only from skeletal remains found on the Channel Islands off southern California. Its disappearance might have been caused by the early Indians, but a more likely reason is intensive overgrazing of San Miguel Island, the last place where it is known to have survived, by sheep in the 1860s. The next case of a vertebrate succumbing to habitat loss was that of the Gull Island vole, and it is a more precisely documented story. This small mammal, discovered in 1889, lived only on Great Gull Island, located off the eastern end of Long Island, New York. This small 17-acre island of glacial moraine was also prime habitat for a variety of seabirds. When, in 1897, the federal government built Fort Michie, it cleared the island of

Figure 4.5
This dry spring in the area of Comanche Springs, Texas, was once inhabited by the
endangered Comanche Springs pupfish. Photograph by Jim Williams.

Figure 4.6
The major remaining habitat of the Comanche Springs pupfish is limited to short
segments of spring outflow in irrigation canals. Photograph by Jim Williams.

Figure 4.7
Excessive pumping of springs, especially in the arid Southwestern United States, can have a dramatically destructive impact on the flora and fauna of the surrounding area. San Felipe Spring, Del Rio, Val Verde Country, Texas. Photograph by Jim Williams.

Figure 4.8
Clear-cutting of forest land eliminates wildlife habitat. After heavy rains the runoff from these areas carries heavy silt loads into streams, adversely affecting aquatic ecosystems. Northeast of Centerville, Bibb County, Alabama. Photograph by Jim Williams.

Figure 4.9
Stream channelizing destroys aquatic and riparian habitats that can take decades to re-
cover, if they recover at all. North of Luxapalila Creek near Columbus, Mississippi.
Photograph by Jim Williams.

Figure 4.10
Gravel dredging within a stream's channel produces a large silt plume, reducing bio-
logical productivity downstream. Tombigbee River, near Columbus, Mississippi.
Photograph by Jim Williams.

Figure 4.11
Strip mining and the resulting drainage drastically alter the terrestrial and aquatic environment. Northeast of Centerville, Bibb County, Alabama. Photograph by Jim Williams.

vegetation, graded it, and subsequently covered much of the area with cement. In 1898, biologists returned to the island but were unable to find any trace of the Gull Island vole.

While the giant deer mouse, Gull Island vole, and other terrestrial species were losing their habitat during the late 1800s, aquatic species were also experiencing declining populations because of habitat disruption. One such species was the harelip sucker (*Lagochila lacera*), once widespread in clear streams of the east-central portion of the United States. This unusual fish had a peculiar mouth and lips, a characteristic that gave the sucker its common name and its scientific name, *Lagochila,* which means "having a harelip." After its discovery in northern Georgia and Tennessee in 1876, it was reported in Alabama, Arkansas, Indiana, Kentucky, Ohio, and Virginia.

Based on distribution of other fishes, it is likely that the harelip sucker also occurred in Illinois, Missouri, Pennsylvania, and West Virginia.

Although the range of the harelip sucker was the most extensive of recently extinct fishes, its habitat was somewhat restricted. It inhabited pools in clear-water streams with moderate gradient, 50–100 feet wide, and about 3000 feet deep, over a rocky or gravel bottom. During the late 1800s, fishermen in Alabama, Georgia, and Tennessee considered this fish to be "the commonest and most valued species of sucker found in the region" (Jordan and Brayton, p. 280). It was also abundant in the Scioto and Olentangy rivers in Ohio, where fishermen called it may sucker because of the large spawning runs during the month of May. Large adults were 18 inches long and weighed several pounds. Despite its extensive range and abundance, the harelip sucker appears to have been the first North American fish to become extinct in recent times. The last known specimens were collected in the Maumee River in northwestern Ohio in 1893.

Although the early extinction of the harelip sucker did not allow the opportunity for detailed life history studies, at least some biological information has been gleaned from the approximately thirty preserved specimens in museums. The unusual modification of the lips and mouth suggests that the harelip sucker was a highly specialized feeder. Examination of stomach contents revealed that the harelip sucker fed extensively on snails, limpets, fingernail clams, and small crustaceans. Anatomical studies indicate that these fishes probably relied on sight to feed. Clear waters would have been important for the welfare of its prey and would have allowed the sucker to find its prey.

We do not know the actual date of extinction for the harelip sucker, but it was probably during the early 1900s. Its demise had likely begun about a century before, with the cutting of forest and the clearing of land for cultivation. The absence of ground cover vegetation and poor agricultural practices resulted in the loss of soil and a significant increase in the silt load in streams. Streams that formerly ran clear, even after heavy rainfall, became cloudy and turbid. The increased levels of silt and turbidity not only smothered the mollusks and crustaceans, the harelip sucker's most important food, but also reduced visibility, making sight feeding difficult if not impossible. Deterioration of water quality from untreated industrial and municipal waste probably hastened the extinction of the harelip sucker in some areas as well. Scientific literature first noted the extinction of the harelip sucker in the 1940s. Since that time biologists have looked in the most undisturbed, remote streams hoping to locate a remnant population, but none has been found.

Among the many questions the extinction of the harelip sucker raises is how many species adapted to large stream, clear-water habitats did we lose before they were discovered? The most diverse freshwater mollusk fauna in the world is found in North America, and most of it is restricted to the central and eastern United States. Siltation of streams from the clearing of land in the 1800s and early 1900s, the cause of the harelip sucker's extinction, also had a damaging impact on the aquatic mollusk fauna. Ohio naturalists noted declining mollusk populations, resulting from siltation and reduced stream flows, as early as 1858. The problem of stream siltation and pollution was further complicated in the early 1900s with the construction of dams on large rivers. To date, the combined effects of siltation, pollution, channelization, and damming have caused the extinction of at least twelve species of freshwater mussels (figure 4.12). These twelve mussels used to inhabit shallow water and gravel and rocky shoals of large streams in the Ohio, Cumberland, and Tennessee River drainage basins. It is likely that a number of aquatic snails in these rivers are extinct, but the lack of recent survey data makes it impossible to account for these species at this time.

A large dam (figure 4.13) on the Rio Grande caused the extinction of the Amistad gambusia. This small fish was known only from Goodenough Spring and its outflows in Val Verde County, Texas. Its entire habitat

Figure 4.12
The Wabash riffleshell is one of more than a dozen species of freshwater mussels that have become extinct in the past fifty years. Pollution, siltation, dams, and channelization destroyed their habitat. Photograph by Jim Williams.

Figure 4.13
Amistad Dam on the Rio Grande inundated Goodenough Spring, destroying the
only known habitat of the Amistad gambusia. Photograph by Jim Williams.

was inundated by Amistad Reservoir in 1968. As the backwaters of the
reservoir were flooding its habitat, biologists collected a few individuals to
reintroduce them when suitable habitat could be located. Unfortunately,
these efforts failed, and the gambusia was extinct by 1977.

Multiple Factor Extinctions

In some cases the decline and ultimately the extinction of a species involve
a combination of factors, such as direct take, habitat disruption, and introduction of exotic species. Extinction of the silver trout (figure 4.14), a
species closely related to the brook trout, was caused by a combination of
excessive fishing, competition, and predation from exotic fishes. The silver
trout was restricted to two widely separated lakes, Dublin Pond and
Christine Lake, in the Connecticut River system in western New Hampshire. During most of the year, the silver trout inhabited the deeper portions
of these lakes, but it moved to shallow waters in May and June to feed on
emerging aquatic insects and again in October when it spawned. People
noticed a population decline in the late 1800s, and they attributed it to
overfishing, especially when the trout were in shallow water. As the silver
trout declined, several exotic fishes, including Atlantic salmon, rainbow

Figure 4.14
Excessive fishing pressure, competition from exotic fishes, and hybridization with introduced brook trout caused the extinction of the silver trout, which had inhabited two lakes in western New Hampshire.

trout, lake trout, Chinook salmon, yellow perch, and white perch, were introduced, presumably to maintain the fishery in these lakes. In addition to the problem of overfishing and exotics, hybridization with hatchery-raised brook trout might have contributed to the extinction of the silver trout.

On a much larger scale, the demise of four fishes endemic to the Great Lakes is one of the best documented cases of multiple factors causing extinction. The four fishes—blackfin cisco (figure 4.15), deepwater cisco, longjaw cisco, and blue pike—were part of the Great Lakes commercial fishery, which began operating in the early 1800s. Commercial fishing in the Great Lakes grew rapidly during the 1800s, with catches increasing at an average rate of about 20 percent a year through 1890. During this same period, the human population of the Great Lakes area was increasing as well, bringing with it the problems of pollution, siltation, drainage of marshes and wetlands, and damaging of tributary streams. Compounding the pressures on these fishes were the arrival and establishment in the 1900s of several exotic fishes, such as the sea lamprey, alewife, and rainbow smelt; these fishes also contributed to the decline of the Great Lake fishery. The combined effect of overfishing, habitat deterioration, and exotic fishes took its toll on the entire fishery, culminating in the extinction of the above four fishes between 1950 and 1975.

Historically, the blue pike inhabited the deep, clear cool waters of central and eastern Lake Erie, the Niagara River, and western and southern Lake Ontario. During the fall and winter, it shifted to shallower nearshore waters as these areas cooled and became less turbid. The habitat of the blue

Figure 4.15
Extinction of the blackfin cisco was caused by a combination of pollution of its habitat, commercial exploitation, introduction of exotic fishes, and parasitism from the sea lamprey.

pike contrasts sharply with that of the closely related walleye, found in the more turbid, shallower lake areas and streams. Both the blue pike and the walleye were highly sought after commercial and sport fishes in Lakes Erie and Ontario.

The commercial and sport fishery for blue pike was particularly wide reaching and intense. Between 1885 and 1962, fishermen landed approximately one billion pounds. Blue pike was most important in the Lake Erie fishery, where it made up more than 25 percent of the commercial fishery between 1915 and 1959. In some years, it exceeded 50 percent of the total commercial fishery in Lake Erie. Though the blue pike commercial fishery was less important in Lake Ontario, averaging about one-tenth the production of Lake Erie, the sport fishery was important in both Lakes Erie and Ontario. In addition to sport fishing from private boats, there was a large charter boat fishery until the blue pike population crashed in the late 1950s.

The decline and ultimate extinction of the blue pike were extremely rapid for such an abundant and widespread species. After the population crashed in 1958, it was still reported in the fishery landings, at low levels, until 1965, and occasional individuals were reported until 1970. The precise mixture of stresses that brought about extinction is still open to speculation. There is little doubt, however, that the intense fishery, pollution, siltation, and predation by exotics were among the causes. Like the blue pike, the blackfin, deepwater, and longjaw ciscos were all subjected to heavy commercial fishing pressure. Subsequent pollution of their habitat and the invasion of exotic fishes proved to be deadly.

Island Extinctions

If fishes restricted to inland bodies of water have proved especially vulnerable to human environmental disruptions, so have terrestrial organisms that live isolated on islands. Puerto Rico, for example, seems to have lost at least ten vertebrates, all by the year 1900. No part of the United States, however, has suffered as high a rate of extinction as the Hawaiian Islands. Of the approximately seventy known endemic species and subspecies of birds known from Hawaii, twenty-four are now extinct and thirty are endangered or threatened.

Although the arrival of the Polynesians and their domestic animals could have caused some problems for the birds, the most dramatic decline began shortly after the visit of Captain Cook in 1778. European sailors and settlers released cattle, sheep, pigs, goats, horses, and rabbits, which proliferated and devastated the fragile virgin forests. Much native habitat was also cleared for agriculture. Other introduced species included predatory cats, rats, and mongooses, which were especially hard on the flightless species of birds that had evolved in Hawaii. A tropical mosquito (*Culex quinque-fasciatus*) was also introduced. It acted as a carrier for the spread of avian malaria and other diseases and is thereby thought to have caused the disappearance of some Hawaiian birds.

Mainland Habitat Disruption

Mainland birdlife has not escaped the effects of habitat disruption. An example is the dusky seaside sparrow, with the most restricted range of all North American birds. It was discovered in the St. Johns River Valley west of Titusville, Florida, by Charles Maynard in 1872. He observed only a single bird, which he described as a black and white shore finch. It was subsequently found to be abundant in the cordgrass (*Spartina bakerii*) savannas and marshes of the Indian River on Merritt Island, Brevard County, located in east-central Florida. Through the 1940s the dusky was abundant on Merritt Island, where an estimated 2000 pairs lived on approximately 6000 acres of marsh habitat. A survey of the dusky in marshes on the St. Johns River in 1968 revealed nearly 894 males.

The ideal nesting and wintering habitat of the dusky seaside sparrow was a wet or moist cordgrass savanna with scattered cabbage palms (*Sabal palmetto*), hammocks, and ponds. This habitat typically occurred at an elevation of 10–15 feet above mean sea level. Areas below this elevation were usually wetter, with denser vegetation, and the areas above the 15-foot elevation were drier, with little cordgrass.

During the 1950s, as the upland areas of Merritt Island were being cleared for what is now the Kennedy Space Center, the workers encountered a serious mosquito problem. When they sprayed, diked, and flooded the marshes on Merritt Island in an effort to control the mosquitos, continuous inundation of marshland habitat destroyed the cordgrass savannas, thus making the area unfit for the dusky. Diking and spraying the marshes on Merritt Island were the beginning of the downfall of the dusky. As the cordgrass habitat began to change, the dusky population started its decline, and by the mid-1970s only a few pairs remained. In a last ditch effort, one of the dikes was opened to restore natural water movement. Even though the native vegetation gradually returned to some areas, it came too late for the dusky population on Merritt Island.

Dusky habitat along the St. Johns River disappeared gradually, caused by a variety of habitat alterations. As people invaded the dusky territory during the late 1950s and the 1960s, the area was drained, cleared, and converted to pastures and residential developments. As the habitat continued to shrink, developers burned the remaining savannas and forced the resident duskies out of the area. As their living space disappeared, the birds disappeared along with it. Between 1968 and 1977 the dusky population declined from almost 900 pairs to about 30.

In 1979 and 1980, biologists searched the remaining habitat along the St. Johns Valley and Merritt Island in hopes of finding duskies for a captive program. They found only seven, and all of them were males. They took six into captivity and tried to crossbreed them with a closely related subspecies of sparrow. Biologists were hopeful that they could maintain a large portion of the dusky gene pool by cross breeding the dusky and then back-crossing the offspring with pure duskies. Thus far, however, these efforts have been unsuccessful, and time is fast running out. Of the six males originally taken into captivity, only three remain, and as this book goes to press, only a single pure dusky seaside sparrow survives.

Prevention of Human-Induced Extinction

Protecting Species
To prevent extinction of a species or a population the first requirement is to understand the factors causing the downward trend, and the second is to control these factors. The task of identifying factors that threaten a species with extinction is a relatively easy job. A team of biologists can accomplish

this in one or two years. Controlling these factors, however, is frequently a difficult and never-ending job. We know, for example, the conditions that threaten most endangered species with extinction, but preventing extinction and recovering many of these species requires a considerable amount of time, effort, and money.

Hunting for food, fur, and feathers and protection of crops and live-stock were the principal causes of the first wave of modern animal extinction across America. Although this killing proceeded virtually unchecked for decades, it eventually attracted enough attention in the early 1900s to warrant corrective action. This came in the form of federal and state laws to protect animals from unregulated take. Game laws that imposed bag limits and restricted or closed hunting seasons probably saved many species from extinction.

Protecting Habitats

As game laws were being enacted to regulate direct take of fish and other wildlife, it was becoming obvious that the deterioration and loss of habitat were also involved in the decline and extinction of wildlife. The first federal act that indirectly resulted in habitat protection was the establishment of Yellowstone National Park in 1872. This act provided "a public park or pleasuring-ground for the benefit and enjoyment of people . . . their retention in their natural condition . . . against the wanton destruction of the fish and game." Although there was some initial confusion over the extent of wildlife and its habitat in Yellowstone National Park, this was eliminated in 1894 with the passage of the Yellowstone Park Protection Act. This legislation established without question the first refuge where wildlife and its habitat received complete and total protection from human intrusion.

President Theodore Roosevelt established by executive order in 1903 the first wildlife refuge—Pelican Island, off the east coast of Florida. Subsequently, from 1905 to 1906, Congress enacted legislation that authorized a federal refuge system for fish and wildlife. This was the first effort by the federal government to set aside habitat specifically for the protection of fish and wildlife. To date, the U.S. Fish and Wildlife Refuge System has grown to approximately 380 refuges, covering about 34 million acres of land and water. Other lands controlled by federal agencies, such as the Forest Service and the Bureau of Land Management, provide important habitats for a variety of plants and animals. The primary management for most of these lands, however, is for resources other than fish and wildlife habitats.

Today the most important means to prevent extinction of fish, wildlife, and plants is the protection of their habitats. Ash Meadows, a desert wetland located in southern Nevada, is a good example. Ash Meadows includes almost 50,000 acres of spring-fed wetlands and arid uplands and is located along the Nevada-California border, about 90 miles northwest of Las Vegas and 40 miles east of Death Valley. The area harbors more than two dozen plants and animals found nowhere else in the world. This is the largest concentration of endemic plants and animals in such a small area in the United States. It is also a desert oasis that has long attracted the attention of humans. Abundant relics indicate the presence of Paiute Indians about 9000 to 11,000 years ago. Abandoned homesteads dating from the early 1900s indicate more recent attempts to settle the area. It was not until the 1950s and 1960s that human presence began to affect adversely the Ash Meadows area.

During the 1950s, exotic animals, including bullfrogs, crayfish, and mosquito fish, had become established in several springs in Ash Meadows. By the late 1950s, one species, the Ash Meadows killifish, was extinct. And several other endemic fishes were declining, while the exotic species increased. The 1960s witnessed the beginning of major physical changes in Ash Meadows, as a large farming operation came to the area. With cultivation of the land and installation of an extensive irrigation system, the water level in the springs decreased. This activity was well on its way toward causing the extinction of the endangered Devil's Hole pupfish and several other endemic species until a Supreme Court decision in 1976 limited the withdrawal of groundwater in Ash Meadows. This court decision forced the farming operation to close.

In 1980, a land development company bought the farming operation's approximately 13,000 acres of land and promptly announced plans for the construction of its first 4000-lot subdivision, one of several it intended to develop in Ash Meadows. With the construction of roads and the clearing of land adjacent to springs, several of the endangered plants and animals were pushed to the brink of extinction. Fortunately for the wildlife, these activities allowed conservationists to invoke the protective measures of the Endangered Species Act of 1973. And once the destructive actions of the developers had been halted, The Nature Conservancy, a national conservation organization, began discussions with the owning company to purchase its land in Ash Meadows. In 1984, The Nature Conservancy purchased the 12,614 acres of Ash Meadows held by the developer and subsequently sold it to the U.S. Fish and Wildlife Service. This land plus additional public domain lands in the area are now part of the recently established Ash Meadows National Wildlife Refuge. After thirty-five years of abuse, Ash Meadows and its endemic plant and animal life appear to have been spared. In addition to protecting the endemic plants and animals, refuge plans call for the reestablishment of several of the original marshes and desert wetlands drained in past years. These areas will once again provide habitats for a variety of resident and migratory wildlife.

Captive Propagation

Although habitat acquisition and protection as a park or refuge is desirable, it is not always possible. In the coming years it may take more than habitat protection alone to prevent extinction of some species. Activities such as captive propagation, habitat renovation, and species reintroduction will be necessary to maintain effective breeding populations of some species. Captive propagation is presently being used as a temporary expedient to prevent extinction of some species. It is a short-term solution to a long-term problem. A species can be maintained in captivity until a secure habitat is found or while a former habitat is being renovated. Captive breeding can also be used to produce a large number of individuals to supplement existing populations. Long-term maintenance of a species in captivity is not biologically or economically sound, and it should be avoided if at all possible.

Habitat Restoration

Fish and wildlife managers have been restoring habitats for many years. In the past they have been concerned primarily with terrestrial wildlife. Practices such as planting ground cover and trees on abandoned farm lands and restoration of marshes and wetlands have been beneficial to a variety of wildlife. Today, as loss of habitat backs more and more fishes and other aquatic species into the corner of extinction, this tool may be the only viable option to save some aquatic organisms. Renovations of aquatic habitats include silt removal, restoration of riparian vegetation, diverting channelized streams into their original channels, reclamation of coal mines to prevent acid mine drainage, and removal of dams. Such efforts are already underway, albeit on a limited scale. Equipment for removing silt from gravel-bottomed streams, for example, has been developed and field-tested in several streams in the Pacific Northwest in an effort to restore spawning areas for salmon. A renovation project underway on twelve miles of the channelized Kissimmee River in Florida will divert this stream back into its original meandering course, thereby restoring adjacent marshlands and oxbow lakes. Projects like these can help save species pushed to the brink of extinction and, at the same time, prevent others from coming under the threat of extinction.

Control of Exotics Introduction

Biological pollution, the introduction of exotic or nonnative organisms, has long been recognized as a serious problem for many ecosystems and the life they contain. Exotic organisms have contributed to the demise of half (twelve of twenty-four) of the extinct U.S. fishes. Exotics are also considered a threat to twenty-nine of the fifty-four fishes presently listed as threatened or endangered by the U.S. Fish and Wildlife Service. Federal and state laws and regulations control importation and release of exotic organisms. Unfortunately these laws and regulations are not comprehensive enough or enforced strongly enough to prevent the deliberate or accidental release of exotics. There is a general lack of public understanding about the potentially disastrous effects of releasing exotic organisms. Evidently there is a similar lack of understanding among many agricultural, fish, and wildlife scientists, who continue to release exotics without knowing or investigating the implications of their actions. Examination of a few prominent examples of intentionally or accidentally introduced exotics, such as the English sparrow, starling, Norway rat, toad, carp, Asiatic clam, and Japanese chestnut, make a convincing case against release of any exotic organism (figure 4.16).

The problem of exotic organisms requires simultaneous action on two fronts: (1) elimination or control of these species that have been introduced and have become established, and (2) limiting or preventing future introductions. Responsible state and federal agencies will require considerable time and effort to accomplish these tasks. Unless measures are taken to control established nonnative organisms, we will continue to see the extinction of our native fish and wildlife by these alien organisms.

We started this chapter with a hypothesis. Because the temperate "backyard" has a greater number of individuals within a species and because they are spread over an extensive range, temperate wildlife should be less vulnerable to the forces of extinction. The true picture has proved otherwise— there simply is not any evidence that the temperate world is particularly more secure in its wildlife diversity than any other corner of the planet. The stuffed specimen of the last passenger pigeon, which once flocked by the billions in the forests of North America, stands as stark witness to

Figure 4.16
Bear Lake in Rocky Mountain National Park is one of several sites where exotic trout have been removed so that native greenback cutthroat trout, a threatened species, could be reintroduced. Photograph by Jim Williams.

temperate vulnerability. In short, we do not have to look to the tropics for endangered species; mass extinction holds court just about everywhere.

Although extinctions will continue to occur in temperate regions, including North America, a public policy has been developing in the United States to slow or stop this process. This policy began with establishment of wildlife conservation laws around 1900 and culminated in the passage of endangered species conservation legislation in 1966 and 1969 and of the Endangered Species Act of 1973. The purposes of the 1973 act are to provide a means of conserving the ecosystems on which endangered and threatened species depend and to provide a program for the conservation of such species. In passing this act, Congress found that various species of fish, wildlife, and plants in the United States had been driven to extinction because economic growth and development had been untempered by adequate concern and conservation. It was also noted that other species of fish, wildlife, and plants had been so depleted that they, too, were in danger of extinction. Congress recognized that these species of plants and wildlife are of aesthetic, ecological, historical, recreational, and scientific value to the United States and pledged to conserve them to the extent practicable.

To achieve the purposes of the Endangered Species Act of 1973, Congress encouraged the states and other interested parties to develop conservation programs that would meet national and international standards. Federal financial assistance and other cooperative efforts provided the incentive for states and other parties to develop such programs. As a sovereign nation in the world community, the United States also pledged, through conservation treaties and international agreements with Canada, Mexico, and other foreign countries, to support the conservation of fish, wildlife, and plants facing extinction. Congress further declared the policy "that all Federal departments and agencies shall seek to conserve endangered species and threatened species and shall utilize their authorities in furtherance of the purposes of this Act" (Endangered Species Act of 1973, Public Law 93205, Dec. 28, 1973, p. 2).

Twenty years have gone by since the passage of the Endangered Species Preservation Act of 1966, the first of the three endangered species laws. Has the endangered species legislation worked, and did it prevent the extinction of some species? The answer is an unqualified yes! Not only has it prevented the extinction of some kinds of animals and plants, it has slowed or reversed the downward trend of many others. While we have lost some species during the past twenty years, most of these were beyond the point of no return before the passage of endangered species legislation.

In closing we would like to respond to the frequently asked question, "How can I help?" or "What can I do?" We would encourage you to become active in conservation organizations at the local, state, and national level. The efforts of these organizations are important; they educate the public, monitor actions of state and federal agencies, raise funds for the purchase of fish and wildlife habitat, and effect changes in public policy on conservation issues. The more you care about fish and wildlife today, the more fish and wildlife there will be to care about tomorrow.

Table 4.1.
Extinct vertebrates of the United States, U.S. territories, and Canada since 1492

Common name	Scientific name	State or region	Date
Fishes			
Miller Lake lamprey	*Lampetra minima*	Oregon	1953
Longjaw cisco	*Coregonus alpenae*	Great Lakes: Lakes Erie, Huron, Michigan	1970s
Deepwater cisco	*Coregonus johannae*	Great Lakes: Lakes Huron, Michigan	1950s
Blackfin cisco	*Coregonus nigripinnis*	Great Lakes: Lakes Huron, Michigan, Ontario, Superior	1960s
Yellowfin cutthroat trout	*Salmo clarki macdonaldi*	Colorado	1910
Silver trout	*Salvelinus agassizi*	New Hampshire	1930s
Thicktail chub	*Gila crassicauda*	California	1957
Pahrangat spinedace	*Lepidomeda altivelis*	Nevada	1940
Phantom shiner	*Notropis orca*	New Mexico, Texas, Mexico	1975
Bluntnose shiner	*Notropis simus simus*	New Mexico, Texas	1964
Clear Lake splittail	*Pogonichthys ciscoides*	California	1970
Las Vegas dace	*Rhinichthys deaconi*	Nevada	1950s
June sucker	*Chasmistes liorus liorus*	Utah	1935
Snake River sucker	*Chasmistes muriei*	Wyoming	1928
Harelip sucker	*Lagochila lacera*	Alabama, Arkansas, Georgia, Indiana, Kentucky, Ohio, Tennessee, Virginia	1900

Table 4.1 (continued)

Common name	Scientific name	State or region	Date
Tecopa pupfish	*Cyprinodon nevadensis calidae*	California	1942
Shoshone pupfish	*Cyprinodon nevadensis shoshone*	California	1966
Raycraft Ranch killifish	*Empetrichthys latos concavus*	Nevada	1960
Pahrump Ranch killifish	*Empetrichthys latos pahrump*	Nevada	1956
Ash Meadows killifish	*Empetrichthys merriami*	Nevada	1957
Whiteline topminnow	*Fundulus albolineatus*	Alabama	1900
Amistad gambusia	*Gambusia amistadensis*	Texas	1977
Blue pike	*Stizostedion vitreum glaucum*	Great Lakes: Lakes Erie, Ontario	1971
Utah Lake sculpin	*Cottus echninatus*	Utah	1928
Lake Ontario kiyi	*Coregonus kiyi orientalis*	New York, Ontario	1967
Alvord cuthroat trout	*Oncorhynchus clarki* ssp.	Nevada, Oregon	1940
Maravillas red shiner	*Cyprinella lutrensis blairi*	Texas	1960
Independence Valley tui chub	*Gila bicolor isolata*	Nevada	1970
Banff longnose dace	*Rhinichthys cataractae smithi*	Alberta	1982
Grass Valley speckled dace	*Rhinichthys osculus relinquus*	Nevada	1950
San Marcos gambusia	*Gambusia georgei*	Texas	1983
Amphibians			
Relict leopard frog	*Rana onca*	Arizona, Nevada, Utah	1960
Golden coqui	*Eleutherodactylus jasperi*	Puerto Rico	1980s
Web-footed coqui	*Eleutherodactylus karlschmidti*	Puerto Rico	1980s
Reptiles			
Navassa iguana	*Cyclura cornuta onchiopsis*	Navassa Island, West Indies	1800s
Iguana	*Leiocephalus eremitus*	Navassa Island, West Indies	1800s
St. Croix racer	*Alsophis sancticrucis*	St. Croix, U.S. Virgin Islands	1900s

Table 4.1 (continued)

Common name	Scientific name	State or region	Date
Birds			
Labrador duck	*Camptorhynchus labradorium*	Northeastern United States, southeastern Canada	1878
Heath hen	*Tympanuchus cupido cupido*	Eastern United States	1932
Kusaie crake	*Aphanolimnas monasa*	Caroline Islands	1828
Laysan rail	*Porzanula palmeri*	Hawaii	1944
Hawaiian brown rail	*Pennula millsi*	Hawaii	1964
Hawaiian spotted rail	*Pennula sandwichensis*	Hawaii	1893
Wake island rail	*Rallus wakensis*	Wake Island	1945
Great auk	*Pinguinus impennis*	North Atlantic	1844
Passenger pigeon	*Ectopistes migratorius*	Central and eastern North America	1914
Culebra Puerto Rican parrot	*Amazona vituta gracelipes*	Culebra Island	1899
Mauge's parakeet	*Aratinga chloroptera maugei*	Puerto Rico	1892
Carolina parakeet	*Conuropsis carolinensis carolinensis*	Southeast United States	1914
Louisiana parakeet	*Conuropsis carolinensis ludoviciana*	South-central United States	1912
Virgin Islands screech owl	*Otus nudipes newtoni*	Puerto Rico, Virgin Islands	1980
San Clemente Bewick's wren	*Thryomane bewickii leucophrys*	California	1927
Lanai thrush	*Phaeornis obscurus lanaiensis*	Hawaii	1931
Oahu thrush	*Phaeornis obscurus oahensis*	Hawaii	1825
Laysan millerbird	*Acrocephalus familiaris familiaris*	Hawaii	1923
Kioea	*Chaetoptila angustipluma*	Hawaii	1859
Oahu oo	*Moho apicalis*	Hawaii	1837
Molokai oo	*Moho bishopi*	Hawaii	1915
Hawaii oo	*Moho nobilis*	Hawaii	1934
Santa Barbara song sparrow	*Melospiza melodia graminea*	California	1967
Texas Henslow's sparrow	*Passerherbulus henslowii houstonensis*	Texas	1983

Table 4.1 (continued)

Common name	Scientific name	State or region	Date
Laysan apapane	*Himatione sanguinea freethi*	Hawaii	1923
Hawaiian mamo	*Drepanis pacifica*	Hawaii	1898
Black mamo	*Drepanis funerea*	Hawaii	1907
Lanai akialoa	*Hemignathus obscurus lanaiensis*	Hawaii	1894
Oahu akialoa	*Hemignathus obscurus ellisianus*	Hawaii	1837
Hawaii akialoa	*Hemignathus obscurus obscurus*	Hawaii	1895
Oahu nukupu'u	*Hemignathus lucidus lucidus*	Hawaii	1860
Oahu akepa	*Loxops coccinea rufa*	Hawaii	1893
Greater amakihi	*Viridonia sagittirostris*	Hawaii	1900
Lanai creeper	*Paroreomyza maculata montana*	Hawaii	1937
Ula-ai-hawane	*Ciridops anna*	Hawaii	1892
Greater Kona finch	*Psittirostra palmeri*	Hawaii	1896
Lesser Kona finch	*Psittirostra flaviceps*	Hawaii	1891
Kona finch	*Psittirostra kona*	Hawaii	1894
Kusaie starling	*Aplonis corvina*	Caroline Islands	1828
Dusky seaside sparrow	*Ammodramus maritimus mirabilis*	Florida	1987
Amak song sparrow	*Melospiza melidia amaka*	Alaska	1980
Mammals			
Puerto Rican shrew	*Nesophontes edithae*	Puerto Rico	1500
Puerto Rican long-nosed bat	*Monophyllus plethodon frater*	Puerto Rico	1900?
Puerto Rican long-tongued bat	*Phyllonycteris major*	Puerto Rico	1900?
Puerto Rican ground sloth	*Acratocnus odontrigonus*	Puerto Rico	1500
Penasco chipmunk	*Eutamias minimus atristriatus*	New Mexico	1980
Tacoma pocket gopher	*Thomomys mazama tacomensis*	Washington	1970
Goff's pocket gopher	*Geomys pinetis goffi*	Florida	1955
Sherman's pocket gopher	*Geomys fontanelus*	Georgia	1950

Table 4.1 (continued)

Common name	Scientific name	State or region	Date
Pallid beach mouse	*Peromyscus polionotus decoloratus*	Florida	1946
Giant deer mouse	*Peromyscus nesodytes*	Channel Islands, California	1870
Chadwick Beach cotton-mouth	*Peromyscus gossypinus restrictus*	Florida	1950?
Gull island vole	*Microtus nesophilus*	New York	1898
Louisiana vole	*Microtus ochrogaster ludovicianus*	Louisiana, Texas	1905
Puerto Rican hutia	*Isolobodon portoricensis*	Puerto Rico	1500
Puerto Rican paca	*Elasmodontomys obliquus*	Puerto Rico	1500
Lesser Puerto Rican agouti	*Heteropsomys insulans*	Puerto Rico	1500
Greater Puerto Rican agouti	*Heteropsomys antillensis*	Puerto Rico	1500
Atlantic gray whale	*Eschrichtius gibbosus gibbosus*	Atlantic Coast	1750
Southern California kit fox	*Vulpes macrotis macrotis*	California	1903
Florida red wolf	*Canis rufus floridanus*	Southeastern United States	1925
Texas red wolf	*Canis rufus rufus*	Oklahoma, Texas	1970
Kenai Peninsula wolf	*Canis lupus alces*	Alaska	1910
Newfoundland wolf	*Canis lupus beothucus*	Newfoundland Island	1911
Banks Island wolf	*Canis lupus bernardi*	Banks and Victoria islands	1920
Cascade Mountains wolf	*Canis lupus fuscus*	British Columbia, Oregon, Washington	1940
Northern Rocky Mountains wolf	*Canis lupus irremotus*	Alberta, Idaho, Montana, Oregon, Wyoming	1940
Mogollon Mountains wolf	*Canis lupus mogollonensis*	Arizona, New Mexico	1935
Texas gray wolf	*Canis lupus monstrabilis*	New Mexico, Texas	1942
Great Plains wolf	*Canis lupus nubilus*	Great Plains	1926
Southern Rocky Mountains wolf	*Canis lupus youngi*	West-central United States	1935
California grizzly bear	*Ursus arctos californicus*	California	1925

Table 4.1 (continued)

Common name	Scientific name	State or region	Date
Sea mink	*Mustela macrodon*	New Brunswick, New England	1890
Wisconsin cougar	*Felis concolor schorgeri*	North-central United States	1925
Caribbean monk seal	*Monachus tropicalis*	Florida, West Indies	1960
Steller's sea cow	*Hydrodamalis stelleri*	Alaska	1768
Eastern elk	*Cervus canadensis canadensis*	Central and eastern North America	1880
Merriam's elk	*Cervus canadensis merriami*	Southwestern United States	1906
Queen Charlotte caribou	*Rangifer tarandus dawsoni*	Queen Charlotte Islands, British Columbia	1935
Badlands bighorn	*Ovis canadensis auduboni*	Montana, Nebraska, North Dakota, South Dakota, Wyoming	1910

Suggested Readings

Glover M. Allen, "Extinct and vanishing mammals of the Western Hemisphere, with the marine species of all the oceans," *Special Publication of the American Committee on International Wildlife Protection* 11. Lancaster, Pennsylvania: The Intelligencer Printing Co., 1942.

Thomas B. Allen, *Vanishing Wildlife of North America*. Washington, D.C.: National Geographic Society, 1974.

R. B. Bury, C. K. Dodd, Jr., and G. M. Fellers, "Conservation of the amphibia of the United States: A review," *Resource Publication* 134, US Department of the Interior, Fish and Wildlife Service. Washington, D.C.: Government Printing Office, 1980.

A. H. Clarke, "Papers on the rare and endangered mollusks of North America," *Malacologia* 10(1) (1970), 1–56.

J. E. Deacon, G. Kobetich, J. D. Williams, and S. Contreras, "Fishes of North America endangered, threatened or of special concern: 1979," *Fisheries* 4(2) (1979), 30–44.

J. C. Greenway, Jr., "Extinct and vanishing birds of the world," *Special Publication of the American Committee on International Wildlife Protection* 13. Lancaster, Pennsylvania: The Intelligencer Printing Co., 1958.

Rene E. Honegger, ed., *IUCN Red Data Book. Vol. 3. Amphibia and Reptilia*. Gland, Switzerland: International Union for the Conservation of Nature, 1979.

W. T. Hornaday, *Our Vanishing Wildlife: Its Extermination and Preservation*. New York: New York Zoological Society, 1913.

David S. Jordan and Alembert W. Brayton, "On *Lagochila*, a new genus of catostomoid fishes," *Proceedings of the Academy of Natural Sciences of Philadelphia* 29 (1877), 280–283.

Warren B. King, *Endangered Birds of the World. The ICBP Bird Red Data Book*. Washington, D.C.: Smithsonian Institution Press, 1981.

D. E. McAllister, B. J. Parker, and P. M. Mckee, "Rare, endangered, and extinct fishes in Canada," *Syllogeus* 54 (1985), 1–192.

R. R. Miller, "Threatened freshwater fishes of the United States," *Transactions of American Fisheries Society* 101(2) (1972), 239–252.

R. R. Miller, *IUCN Red Data Book. Vol. 4. Pisces*. Morges, Switzerland: International Union for the Conservation of Nature, 1977.

G. Nilsson, *The Endangered Species Handbook*. Washington, D.C.: The Animal Welfare Institute, 1983.

Ronald M. Nowak, *Our Endangered Wildlife. Then and Now*. Washington, D.C.: National Parks and Conservation Association, 1982.

Ronald M. Nowak and John L. Paradiso, *Walker's Mammals of the World*, 2 vols. Baltimore, Maryland: Johns Hopkins University Press, 1983.

R. D. Ono, J. D. Williams, and A. Wagner, *Vanishing Fishes of North America*. Washington, D.C.: Stone Wall Press, 1983.

John C. Ralph, "Birds of the forest," *Natural History* 91(12) (1982), 40–45.

Phillip L. Walker, "Archaeological evidence for the recent extinction of three terrestrial mammals on San Miguel Island," In *The California Islands: Proceedings of a Multidisciplinary Symposium,* Dennis M. Power, ed. Santa Barbara, California: Santa Barbara Museum of Natural History, 1980, 703–711.

S. M. Wells, R. M. Pyle, and N. M. Collins, *The IUCN Invertebrate Red Data Book*. Gland, Switzerland: International Union for the Conservation of Nature, 1983.

J. D. Williams and D. K. Finnley, "Our vanishing fishes: Can they be saved?" *Frontiers* 41(4) (1977), 21–32.

Appendix

As the controversy of species conservation versus habitat destruction escalates, so does the number and the rate of human-induced extinctions, taking countless elements of our ever dwindling flora and fauna. Endangered species legislation enacted two decades ago provided the mechanisms for identification and recovery of threatened species. Today, listing species as endangered or threatened continues at a rapid pace. The problem arises with the realization that we are losing species faster than they can be recovered.

While our efforts to recover individual species have been successful in a few cases, it is becoming quite clear that the species approach to recovery for the majority of our threatened flora and fauna is going to be too little too late. Based on our recovery experience during the past two decades, it is clear that, for the majority of species, an ecosystem/watershed recovery program is preferable to the individual species plan.

Habitat loss continues to be the primary problem confronting our threatened flora and fauna. The increased rate of habitat loss is the result of physical, chemical, and biological perturbations, all resulting from human activities. Aquatic species inhabiting our freshwater systems, lakes, rivers, and creeks, are under severe stress, and some indicators suggest that they are on the verge of a major extinction event. Without immediate attention and considerable on-the-ground effort to protect and recover entire watersheds, there is little chance that we will avoid the loss of a significant portion of our freshwater fauna before the end of this century.

Additional Readings

Peggy L. Fiedler and Subodh K. Jain, *Conservation Biology: The Theory and Practice of Nature Conservation, Preservation, and Management,* New York: Chapman and Hall, 1992.

Robert R. Miller, James D. Williams, and Jack E. Williams, Extinctions of North American Fishes During the Past Century, *Fisheries* 1989, 14(6): 22–38.

Matthew H. Nitecki, *Extinctions,* Chicago, Illinois: University of Chicago Press, 1984.

Daniel J. Rohlf, *The Endangered Species Act: A Guide to Its Protections and Implementation,* Stanford, California: Stanford Environmental Law Society, 1989.

Michael E. Soule and Kathryn A. Kohm, *Research Priorities for Conservation Biology,* Washington, D.C.: Island Press, 1989.

U.S. Fish and Wildlife Service, Washington, D.C.: Report to Congress: Endangered and Threatened Species Recovery Program, 1990.

Jack E. Williams, J. E. Johnson, D. A. Hendrickson, S. Contreras-Balderas, J. D. Williams, M. Navarro-Mendoza, D. E. McAllister and J. E. Deacon, Fishes of North America Endangered, Threatened, or of Special Concern: *Fisheries* 1989, 14(6): 2–20.

E. O. Wilson, *Biodiversity,* Washington, D.C.: National Academy of Sciences Press, 1988.

5

Riders of the Last Ark:
The Role of Captive Breeding in
Conservation Strategies

Thomas J. Foose

A flood of destruction is now engulfing the wildlife and wild lands of our planet. Against this flood, zoos and aquariums must provide an ark of captive propagation if many species are to survive through the next few centuries (figure 5.1). The involvement of zoos and aquariums will be especially important for the "charismatic megavertebrates," the large and spectacular species, such as elephants, gorillas, rhinoceroses, crocodiles, and condors. In this chapter we explore the problems, the potentials, and the reality of captive propagation as a component of conservation strategies.

As other chapters in this book have described in some detail, the Earth is now confronted with massive extinctions comparable in scope with other periodic cataclysms in the planet's history, such as those that occurred at the end of the Cretaceous when the dinosaurs disappeared.[1] A major difference between the current crisis and the cataclysms of the past is that the extinctions occurring now are caused by human factors.

We live with the horrifying possibility of a nuclear winter. But we live with the agonizing reality of a "demographic" winter. Michael Soule[2] created this term to describe the foreseeable future when human population growth and development will continue and perhaps intensify its devastation of wild lands, destruction of wildlife, and general disruption of ecological systems on the planet. By most estimates, this demographic winter will persist for at least the next 200 years. At that time, human population should begin to stabilize and probably adjust itself downward, if its does not "crash" before then because of famine, disease, or war.[3,4] Thereafter a period of ecological readjustment, undoubtedly requiring several centuries, will occur and hopefully will result in the restoration of a large amount of habitat available for wildlife.

The problem of the impending cataclysms is quite clear to anyone who observes what is happening. What are the remedies? The general, traditional, and optimal remedy is to protect and thereby preserve some patches of natural habitats and presumably some remnants of their denizens.

Figure 5.1
The zoo ark. Clear problems include the ark's limited capacity, possible overpopulation of some species, and possible destruction of species before they can be taken aboard.

This task is, of course, increasingly formidable as the demographic winter intensifies.

Even if it is possible to save small islands of nature, it may not be enough. Such fragments may not be viable over the long term for genetic and demographic reasons. The problem is that gene pools are being converted into gene puddles as the populations of species are reduced and fragmented. Gene puddles are vulnerable to evaporation in an ecological and evolutionary sense.

The gene pool is the term used to denote the genetic variation or diversity that has evolved in the wild populations of virtually every species. Genetic diversity seems vital to the survival of species at the levels of both populations and individuals. At the population level, genetic diversity seems essential to permit species to adapt to changing or altered environments. At the individual level, genetic variation seems important for health, longevity, and fertility.

Zoos can greatly assist in the preservation of the gene pool. The primary objective of captive propagation, however, is the reinforcement, not replacement, of wild populations. Captive populations are not an end in themselves. Even if the worst occurs in the immediate future, we must be optimistic that in several centuries, perhaps as long as 500 or 1000 years, the human population will stabilize and decline. It should then be possible to restore or even reconstruct ecosystems, but only if we have kept all the parts. Captive propagation is a way to help preserve options for the future.

Two scenarios are relevant here. The first and more dismal one is that a species will disappear entirely from its natural habitat for a while. Moreover, the species may often disappear before the habitat itself does. This is especially the case with the charismatic megavertebrates. There is still much good woolly mammoth and mastodon habitat in the Northern Hemisphere, just no hairy elephants to populate it. In this scenario, zoos are literally an ark that must carry a species until the natural habitat is again tenable.

The more hopeful though no less challenging scenario is that some remnants, the gene puddles, may persist in a natural habitat but not be viable unless reinforced by genetic material from captive propagation. In such a system, both the captive and the wild components are equally important. The wild population permits the species to continue to be subjected to some semblance of natural selection but cannot sustain the genetic diversity necessary for these processes or, ultimately, for the survival of the species. The captive programs can provide a reservoir of genetic variation that can be periodically infused into the natural population or, to use the imagery of G. Evelyn Hutchinson,[5] into the ecological theater so that the evolutionary play can continue.

Thus we are moving toward a world in which survival of many species depends on systems of interactive wild and captive populations that require intensive and similar management. Like it or not, the wild is becoming in many cases a megazoo.

To reiterate, the pervasive problem confronting conservation efforts today is that, when populations become small, they lose diversity quickly. This is true in captivity or in the wild (figure 5.2). And thus it is important to determine the required size of minimum viable populations (MVPs), one of the major buzzwords of conservation biology. MVPs depend on five factors or questions: How much diversity must be preserved and for how long? What is the generation time for a species? How much diversity was there initially? How rapidly does the founder population grow to a minimum viable population?

As for how much diversity must be preserved, the most desirable answer is, of course, all of it. But normally this objective will be impossible or impractical over the next several hundred years until the puddles in which we are trying to preserve biological diversity can be expanded again into pools. In general, it requires twice as large a population to preserve

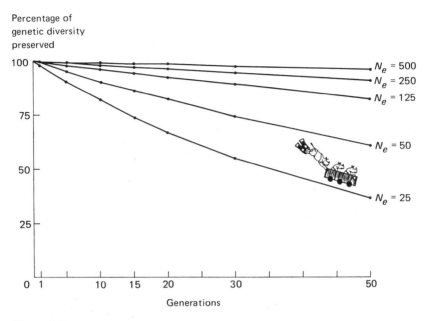

Percentage of
genetic diversity
preserved

Figure 5.2
As the effective size (N_e) of a small population decreases, the genetic diversity also declines rapidly over the generations. Managing populations to maximize N_e is a central problem of captive breeding.

95 percent as it does 90 percent of the diversity. This is complicated when we consider exactly what kind of diversity is significant, that is, maintaining average heterozygosity or retaining as many alleles as possible. The latter generally requires larger populations than the former. Recent analyses have suggested, however, that 90 percent of the average heterozygosity may be a feasible objective for conservation strategies, at least in terms of captive populations and programs (see note 2).

How long must the diversity be preserved? Although forever is the most desirable answer, a more practical objective is until the demographic winter is over; in other words until human population and development equilibrate and allow patches of ecosystems large enough for conservation measures, especially dependence on captive propagation, to be relaxed. Estimates of the demographic winter's duration vary, but it could be as long as a thousand years. A one-thousand-year program may not seem more feasible than forever; thousand-year reichs have had a habit of terminating earlier than expected. Fortunately, the intensive programs to preserve animal genetic diversity, again especially in captive populations, may have a less formidable task.

One amelioration may be that the winter might vary on a species-by-species and area-by-area basis. The ark may be able to land sooner and periodically for some species and areas. Several reintroduction projects, using captive stock of species extinct in the wild, are in progress even now. But these opportunities are likely to be rare and are often transient. Modern Noahs must be prepared to respond to repeated ebbs and flows of the flood.

The most encouraging kind of help on the horizon is reproductive technology. It could help preserve diversity as stored germplasm, that is, sperm, ova, or embryos in a so-called frozen zoo rather than as living animals, which require more space and resources. Unfortunately, these techniques have not been developed for most endangered species, and they will require much further research. Nevertheless, there is some hope that this technology may be generally available for endangered species within 200 years, about the time the human population growth may abate, although still well before the time the planet may return to an ecological equilibrium that once again provides appreciable areas for wildlife.

For these reasons 200 years has recently emerged as a plausible time objective, at least for the captive programs. But whether conditions improve for wildlife in 200 or 500 years, conservation programs are obviously going to require thinking and planning for a longer range than has been the case for humans. This may not be as painful or unprofitable as it might seem in terms of our normal short-term planning and operating horizons. In reality, the initiation phases for such long-term plans are not that difficult or demanding. And, if on the way to the end of the millennium, mankind gets smarter, luckier, or annihilated, the plans can be adjusted for a shorter duration.

The third factor that MVP depends on is how often a population loses its genes, or equivalently, what the generation time of the species is. You can lose your genes only during a reproductive act. Less facetiously, loss of genetic diversity occurs generation by generation (figure 5.2), not year by year. Hence, if population sizes are equal in genetic terms, elephants and golden lion tamarins (a small species of primate from Brazil) would be expected to lose exactly the same amount of genetic diversity over ten generations. Ten generations for elephants, however, is probably 250–300 years; for the tamarins, about 25–50 years. Thus, for a fixed period such as 200 years, the MVP for elephants can be much smaller than it is for the tamarins.

How large was the founder stock or how much diversity was there initially? This fourth factor acknowledges that we cannot preserve more diversity than there is originally. Establishing a protected population, whether it is in zoos or national parks, represents a sampling from the wild

population. The animals that constitute the original population, the founders, might not contain all the genetic diversity that has evolved in the
wild. Without going into technical detail, the smaller the founder number,
the larger the MVP needed for long-term maintenance (figure 5.3). Thus,
usually, the more founders there are, the better. But there is a point of
diminishing returns. For most species, twenty to thirty founders is adequate.
This means that we do not have to decimate wild populations, unless they
are extremely reduced, to establish captive ones. A comparison of figures
5.2 and 5.3 illustrates the significance of founder size for MVP.

Contrary to the most common version of the Noah's ark legend, a
single pair is usually not adequate to found a viable population. In a closer
look at the original texts of the biblical legend, it appears that seven pairs
of animals, not one, were collected for the ark. If true, this is a particularly
relevant statistic, because an average of seven pairs of unrelated species
should contain 90-plus percent of the average heterozygosity of the population from which it is derived.

A final point should be emphasized. The figures cited represent optimal founder numbers. If fewer founders are available, the programs must
compensate. Even if the world were down to its last pair of giant pandas,
we should still try to save them against all odds, genetic or otherwise. It

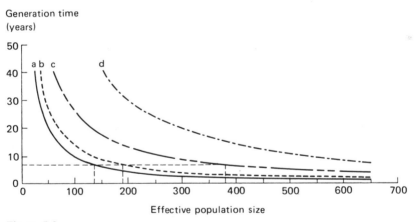

Figure 5.3
Preserving diversity becomes more complex as the number of founders of a population decreases. This graph illustrates the relationship among effective population size,
number of generations, and founder group size required to maintain 90 percent of
original genetic diversity for 200 years. (*a*) No founder effect. (*b*) Twenty founders.
(*c*) Eight founders. (*d*) Six founders. Courtesy of the *International Zoo Handbook*.

would be a bitter irony if species were excluded from the zoo ark because they do not satisfy some optimal standard.

The fifth consideration for MVP is how rapidly the founder population grows to its MVP size. It is important to distinguish founder size from MVP. Founder sizes, particularly for captive populations, can be anywhere from several pairs to a few dozen animals. MVPs sufficient to preserve 90 percent of the average heterozygosity for 200 years are normally several hundred animals. In any case, the faster a population can expand from its founder size to its MVP, the less the diversity that will be lost.

There is yet another extremely important consideration for the determination of MVP sizes. The relevant population size is not merely a total count of all the animals alive. What is critical is the genetically effective size (usually designed N_e). N_e is a measure of the way animals actually contribute their genes to the next generation. Plants and animals are diploid organisms, that is, they carry two versions of each gene. In a population of N diploid organisms ($N = 100$, for example), the gene pool consists of $2N$ (that is, 200) versions of each gene, or simply $2N$ genes. Effective population size is a measure of the expected number of the $2N$ genes in a diploid population of N individuals that will be transmitted to the progeny of the next generation.

An animal that does not reproduce during its lifetime, of course, does not contribute at all to the next generation or to N_e. A male that dominates breeding for a long time may contribute excessively. Prolific mothers were a source of pride for zoos in the days when breeding most exotic species was a novelty and an unusual achievement. Today, in well-managed populations, family planning is the badge of distinction.

Effective population size, then, is a function of the social structure and dynamics of a population. Depending on what these characteristics are, the genetically effective size of a population can vary from a small fraction, one-tenth, for example, to twice the actual number of animals. In natural populations, the social dynamics are such that the genetically effective size is usually less than the total number. In captivity, genetic management is frequently able to produce much higher effective sizes for the population maintained. One of the reasons captive propagation is so important is that zoos and aquariums can maximize N_e, hence the preservation of diversity, from a given number of animals.

There are some who think reintroduction from captivity to the wild will not be feasible even if tenable wild lands are available. One of their major concerns is the disruption these "invaders" might cause to populations in the wild. This concern would apply especially to the second scenario described earlier, in which remnants of the species still survive in their natural habitat. The other major concern about captive propagation is the doubt that animals born and bred in captivity will be able to readapt to a natural environment.

In response to this skepticism and criticism, some limited but encouraging data can be cited from the several projects, past and present, that have attempted reintroductions with some success. The American bison survives today because of captive propagation.[6] When Europeans started to settle the Great Plains of North America, the bison population—about 60 million animals—was deliberately and ruthlessly hunted until only a few scattered herds survived by the end of the nineteenth century—a period of less than seventy-five years. If it were not for stock from the Bronx Zoo and a few private collections that were employed to establish herds in protected refuges, conservationists believe this species would be extinct.

The European model of the bison, the wisent, had at least as rocky a history.[7] A few individuals still inhabited forests in Poland and Russia early in this century, but they were unable to survive the turmoil of World War I and its immediate aftermath. Fortunately, there were animals in zoos that could be propagated, so that now the wisent has been reestablished in what had been its last refuge, the Białowieża Primeval Forest in Poland.

The Arabian oryx[8,9] represents another successful rescue. By the early 1960s, this sandy white species of antelope had been wantonly persecuted to the brink of extinction. In desperation, the World Wildlife Fund (WWF), the Species Survival Commission of the International Union for the Conservation of Nature (IUCN), and the Flora and Fauna Preservation Society (ffPS), organized an expedition to Arabia to capture stock for captive breeding. By the time they arrived, they could obtain only three surviving oryx. A half dozen other animals were "rounded up" from captive collections around the world, however, so that nine founders were eventually assembled to establish a propagation program that has produced in just twenty years a captive population of over 400 animals and, more important, stock that is being reintroduced through exceptionally well-managed projects into Oman (in Arabia), Jordan, and Israel.

A final story apparently on the way to success is the golden lion tamarin, a brightly colored and engaging primate, indigenous to the coastal forests of Brazil.[10] The background of the tamarin is much the same—virtual extermination because of habitat destruction and direct killing. But again, after an uncertain start caused by captive husbandry problems that were solved with concentrated efforts of zoo professionals, a captive population of over 400 animals has developed in roughly twenty years. An intensive program is now in progress to reintroduce captive-born tamarins into protected patches of forest in their native Brazil. Early results indicate that the animals are readapting satisfactorily. An even more encouraging development is that the Brazilian citizenry has become enthusiastically involved in the restoration of this tamarin, and many ranchers are even providing sanctuary on their property for the reintroduced colonies.

Even with the California condor, captive propagation might yet save a species clearly "on the way out."[11,12] Wildlife biologists have been observing the decline of this species for the last half century. Over the last two decades, a conservation program has developed fitfully. Regrettably, conflicts of philosophy, jurisdiction, and sometimes personalities among the conservationists have been the principal reasons for indecision and inaction. Lamentably, the call for help from captive propagation might have occurred too late for this species. Indeed, with only about half a dozen birds surviving in the wild as of 1985, there are still conservationists who oppose or impede captive breeding. There are even advocates of "death with dignity." Despite such resistance, an active program of captive propagation has been initiated and, based on the successful efforts with the closely related Andean condor, the prospects seem moderately good that this relict of the Pleistocene will once again wing its way over the California coast.

These case studies are a basis for optimism that reintroduction of captive-bred animals is more than just a dream. If there are cases for which reintroductions are difficult, however, advances in reproductive technology, such as artificial insemination and embryo transfer, might enable the transfer of genetic material from captive to wild populations without the complications that reintroducing animals might entail. Exploring such possibilities is going to be one of the most exciting areas of wildlife research in the near future.

Despite some skepticism and criticism, most conservationists recognize the importance of captive propagation in conservation strategies. Reciprocally, zoos and aquariums are recognizing their responsibilities for wildlife conservation and, accordingly, are organizing coordinated scientific programs for the propagation of endangered species.

There are two basic problems for programs to propagate species in captivity as part of conservation strategies. The first is selection of taxa for the programs, that is, who gets on the ark. The second is management of the taxa selected as biological populations. It is more informative to discuss the second problem first.

Zoos and aquariums can significantly contribute to conservation only if they manage their collections as biological populations. This entails genetic and demographic management as well as all the husbandry that keeps animals alive and permits or promotes their breeding.

In terms of genetic management, the basic objective, as stated before, is to preserve as much of the inheritable variation or diversity of the wild gene pool as possible for as long as necessary to help the species to survive, ultimately as components of their natural ecosystems. For details of this genetic management, the reader should consult a number of recent summaries on the subject.[13,14] In brief, genetic management attempts to maximize the genetically effective size of the population and to avoid too much inbreeding. Specific practices include (1) equalizing the sex ratios and, more important, the family sizes of the animals that reproduce (the total number of offspring any one animal produces in its lifetime); (2) rectifying disparities in the bloodlines or founders (each founder out of the wild population is considered to establish a different bloodline); and (3) avoiding reproduction between closely related animals. Achieving these objectives requires that matings be precisely specified, breeding closely regulated, and animals frequently moved from one institution to another to produce better genetic combinations.

Genetic management is one part of the story. Demographic management is the other. In the wild, natural populations seem to equilibrate around some long-term average size that ecologists call the carrying capacity of that environment for that species. At least this concept is a convenient if somewhat fictional representation of nature. Various natural regulatory mechanisms stabilize the wild populations around these carrying capacities. In captivity it is also important to stabilize populations around some artificially established carrying capacity. Carrying capacity in captivity is a compromise between a population large enough to be genetically and demographically viable (that is, an MVP) but not too large to deprive other taxa of space on the zoo ark. Fluctuations in captive populations can reduce the genetically effective size and result in population explosions, or population crashes and extinctions. Growth of species whose reproductive husbandry has been mastered can be explosive (figure 5.4). Demographic management in captivity must therefore regulate population explosions, crashes, extinctions, or fluctuations.

Demographic management is not simply a matter of controlling numbers, however. Age distributions are another critical factor. Again, this is not the appropriate place for a technical discussion of demographic management.[15] It is important, however, that captive populations be managed so that there is a regular and healthy replacement of older animals by younger ones. So often in the past (before the days of good population management) zoos reacted to population explosions and the problem of how to accommodate or dispose of animals by curtailing reproduction. The

Number of animals

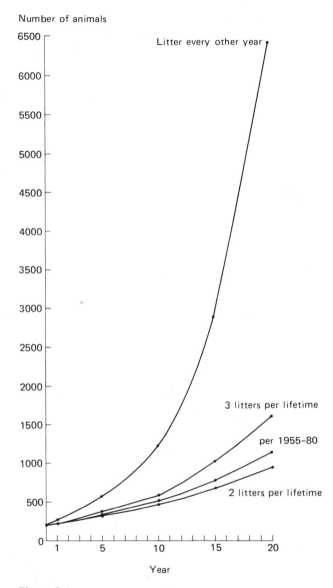

Figure 5.4
Captive populations of Siberian tigers have an explosive potential for growth. Even the projection for litters every other year is nowhere near the full potential of this species, which can produce two or three litters a year.

result was populations with top-heavy age distributions, with too many older and too few younger animals causing the kind of demographic fluctuations described (figure 5.5). Stable demographic management requires a combination of birth control and animal removal. Some of the animals, it is hoped, might be returned to natural habitats.

There are other types of demographic problems in captivity. The gorilla is an eminent example of a species for which captive reproductive husbandry has not yet been mastered. As a consequence, the captive population of gorilla is not self-sustaining and will decline to extinction in about thirty years if the problems are not solved. A particular difficulty seems to be the failure of many captive males to breed after early adulthood, if at all. A few institutions, however, seem to be especially successful in propagating gorillas. Consequently, an intensive research and management program is now underway to both master and transfer the art or science of gorilla breeding.

Formulation of plans for both genetic and demographic management entails prescribing which animals should mate with each other, how many times and at what ages animals should breed, and how many offspring should be produced. Implementation of these scientific plans requires not only a phenomenal amount of cooperation but also coordination. To provide this kind of organization, the American Association of Zoological Parks and Aquariums (AAZPA) has developed the Species Survival Plan, or SSP. The SSP is an attempt to organize the zoos and aquariums in the United States and Canada for the kind of captive propagation programs described.

Each SSP program is organized around a species coordinator, a person with particular interest, experience, or expertise with the animal in question. The species coordinator is assisted by a management committee, known as the propagation group, which is composed of nine members elected from and by the institutions participating in the program. Every major zoo and aquarium in the United States and Canada with specimens of the taxon usually participates in the SSP program. The core of the SSP is the Masterplan for Management and Propagation. This masterplan provides institution-by-institution and animal-by-animal recommendations on mate selection, animal relocations (from one zoo to another to produce better genetic combinations), breeding schedules, and culling programs.

Although the SSP is currently the most advanced of the scientific and coordinated programs for captive propagation of endangered species, there is progress in other parts of the zoo world as well. The British Federation of Zoos has initiated programs similar to the SSP. Eventually, the hope is

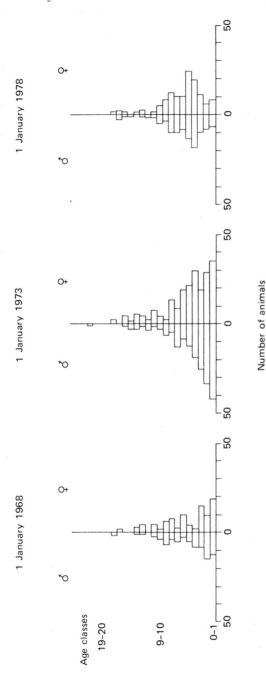

Figure 5.5

This graph of the age structure of Siberian tigers in North America from 1968 to 1978 shows how the age structure of a population can become unstable (top-heavy with older animals) as a result of curtailed reproduction in response to a population explosion.

Figure 5.6
The project director of the International Species Inventory System using the centralized computerized data base so essential to the management and propagation of animals in captivity.

that regional (that is, continental) SSP-type programs will develop worldwide and coordinate in an international system for endangered species
conservation.

All these types of programs are possible only if basic data are available
for analysis. And compilation and analysis of data on the scale required by
SSP programs are feasible only with the aid of computers. Yes, the computer
revolution has also extended to zoos and aquariums. The International
Species Inventory System, or ISIS (the Egyptian goddess of fertility), is a
centralized and computerized data base containing information on most
species in North American zoos and aquariums and, to an ever-increasing
extent, captive facilities worldwide. While this central repository provides
data and analysis for SSP programs, species coordinators are using micro-
computers to perform much of the work themselves (figure 5.6).

Successful propagation of animals in captivity presents a paradox and
a predicament. Good population management will eventually cause every
animal to be surplus for genetic and demographic reasons because each
animal will eventually fulfill its contribution to the next generation and
hence to the survival of the species. It is often difficult for the layperson to
appreciate how there can be a surplus of an endangered species. How to

contend with this surplus and still provide sanctuary for as many taxa as possible is one of most formidable problems confronting the zoo ark.

There are essentially two options for animal surplus: maintenance or disposal. Maintaining surplus reduces the already limited capacity of the zoo ark. Disposal of surplus entails moving the animal elsewhere. Sometimes there are opportunities to return the animal to a natural habitat. Occasionally it is possible to place the animal at another facility. AAZPA zoos, however, operate under a code of ethics that prohibits them from placing animals with unqualified recipients. And to a greater and greater extent, all the qualified recipients have already been recruited for maintenance of nonsurplus animals with the zoo ark.

As a consequence, many zoo professionals believe euthanasia will be essential if the conservation responsibilities of captive facilities are to be fulfilled. It is, however, important to distinguish the designation from the disposal of surplus. The SSP does not issue death warrants. The decision to employ euthanasia for management reasons (as opposed to medical) is currently a matter determined by each community according to its own standards. It is vital, however, for the zoo ark that these communities be as well informed as possible about the conservation implications and consequences of such decisions.

A simple supply and demand analysis shows how critically limited the capacity of the zoo ark is. The demand is enormous and expanding. Conservationists estimate, conservatively I think, that at least 1500 taxa of mammals, birds, reptiles, and amphibians will be so endangered by the middle of the next century that assistance from captive propagation will be essential for their survival (see note 2).

In contrast to the demand, the space and resources available in zoos, the captive habitat, is woefully limited. Foose and Seal[16] have used wild felids as an example to demonstrate that captive facilities can accommodate large populations of only a fraction of the taxa that are, or soon will be, endangered (table 5.1). In an even more global analysis, Conway[17] has estimated that, even if zoos utilize their space and resources as efficiently as possible for endangered species, they will be able to provide help for only about 1000 taxa of terrestrial vertebrates, far short of the need. These calculations indicate why there is such concern about carrying surplus animals.

A general solution to contend with the limited space in the zoo ark is to establish a carrying capacity for every taxon selected. This requires determining what taxa are in need of captive help, assessing what captive habitat space and resources are available (realizing that captive habitat tends

Table 5.1.
Capacity of ISIS captive facilities for larger felids

Species	Existing subspecies	Already endangered subspecies	ISIS facilities with species	Current population in ISIS facilities	Number of subspecies accommodated if the population is		
					100	250	300
Lion	11	1	97	381	4	1	1
Tiger	8	8	110	450	4	2	1
Jaguar (*Panthera onca*)	8	8	65	178	2	1	0
Leopard (*Panthera pardus*)	15	15	72	246	2	1	0
Snow leopard (*Panthera uncia*)	1	1	35	128	1	0	0
Cougar (*Felis concolor*)	29	2	69	173	2	1	0
Clouded leopard (*Neofelis nebulosa*)	4	4	20	63	1	0	0
Cheetah (*Acinonyx jubatus*)	6	6	32	166	1	0	0
Total	82	45		1785	18	6	3

The three groups are lions and tigers, "intermediate" cats, and other large felids. The current population in ISIS facilities is a crude measure of their capacity for these kinds of animals. The last three columns indicate how many subspecies can be accommodated if the carrying capacity is established at the given number of animals of each subspecies.

to be rather species or group specific—elephants and snakes occupy different kinds of enclosures and have dissimilar needs), and allocating the available space and resources to as many taxa as possible.

Within the AAZPA, three basic criteria govern which taxa are being selected for the SSP program: endangerment in the wild, feasibility in captivity, and uniqueness in relation to other taxa. Endangerment in the wild is ultimately the primary criterion. Certainly some cases are more urgent than others; some taxa are still not classified as endangered. But, at least for the megavertebrates, most will be endangered by the middle of the next century if not before.

Feasibility in captivity encompasses a number of considerations: Are there enough animals available for a viable foundation? Are there facilities and staff committed to an intensive management program for the taxon? Can the taxon be kept alive and reproduced in captivity? If not, what are the prospects that a concerted effort will master these skills? A spectacular example of an unfeasible taxon is the blue whale.

Probably the most difficult of the three criteria to apply is uniqueness in relation to other taxa. There are currently six endangered subspecies of tiger. Once one or two subspecies have been selected for the zoo ark, how can we justify adding more kinds of tigers when there are so many other significantly different life forms (lions or jaguars, for example) desperately in need of help? With their limited capacity, the captive programs must incorporate as much of the taxonomic diversity of the planet as possible. Thus species that are the only representative of their genus, family, or order will receive priority over species with many close relatives, or subspecies.

Using these criteria, the SSP has designated thirty-seven taxa (table 5.2) for zoo ark preservation. Hundreds more need to be designated, and the SSP is currently expecting to add about a dozen new taxa a year. Designation of species has been conservative in the early stages to ensure that the programs are operating well before they are extended. Eventually, the objective is to have at least 1000 taxa on the SSP ark.

Despite all these efforts, there is the need to accommodate more taxa on the zoo ark. Moreover, the capacity of the ark normally assumes the minimum viable population. Larger populations are always preferable. It would be safer to preserve tigers with a captive population of 5000 rather than 500. And that is why there are attempts to expand the ark capacity both territorially and technologically.

Territorial expansion is occurring in a number of ways. Zoos and aquariums are becoming larger and more naturalistic. The San Diego Wild Animal Park is a good example. Many institutions are developing spacious

Table 5.2.
Taxa designated for the AAZPA Species Survival Plan (SSP)

Amphibians	Mammals	Asian elephant
Puerto Rican crested toad	Ruffed lemur	Indian rhinoceros
	Black lemur	Sumatran rhinoceros
Reptiles	Golden lion tamarin	Black rhinoceros
Chinese alligator	Lion-tailed macaque	White rhinoceros
Radiated tortoise	Gorilla	Asian wild horse
Aruba Island rattlesnake	Orangutan	Grevy's zebra
Indian python	Asian small-clawed otter	Chacoan peccary
Madagascan ground boa	Maned wolf	Barasingha
	Red wolf	Okapi
Birds	Red panda	Gaur
Bali mynah	Siberian tiger	Arabian oryx
White-naped crane	Asian lion	Scimitar-horned oryx
Andean condor	Snow leopard	
Humboldt's penguin	Cheetah	

facilities solely for propagation and research—the Conservation and Research Center of the National Zoological Park in Front Royal, Virginia, for example. There are also experimental projects in progress for cooperative programs between zoos and private facilities that have both space and money, as well as a sincere interest in conservation. Notable examples are the herds of Grevy's zebra and scimitar-horned oryx on exotic animal ranches in Texas.

Technological expansion is occurring through such reproductive techniques as germplasm collection and storage (the "frozen zoo"), artificial insemination, and embryo transfers. By developing these technologies, we can expand the effective population sizes of a few hundred living animals to literally thousands of potential specimens in the freezers (liquid nitrogen storage tanks). Unfortunately, although the potential of these technologies is great, their reality is remote. Even artificial insemination is available for only a few wild and endangered species. Extensive and expensive research is needed before these techniques can be readily applied to most endangered species.

Zoos are working on the problem. A group of researchers led by Ulysses Seal, the SSP species coordinator for Siberian tigers, is trying to develop artificial insemination for tigers and other large cats. Scientists at the Bronx Zoo have successfully transferred an embryo from a gaur (a large species of wild cattle from Southeast Asia) into a holstein cow surrogate mother. Cross-species transfers across greater taxonomic distances have been achieved with bongo and eland antelopes (figure 5.7). Who knows? Perhaps some day the technology will advance to the point where the woolly mammoth can be "resurrected" from the frozen tissues that have been recovered from the Siberian tundra. Russian scientists have reportedly been working toward this goal for a number of years, although many scientists believe this miracle will never occur.

An appropriate conclusion for our saga of the zoo ark is a case study that describes how captive propagation can integrate with conservation efforts in the field to preserve species. The Sumatran rhinoceros, also known as the hairy rhinoceros and the Asian two-horned rhinoceros, is one of the most endangered species in the wild (figure 5.8). It is one of the five extant species of rhinoceros, all of which are endangered to some degree. Four of these species, the Indian, the Sumatran, the black, and the white, have already been designated for the SSP program.

Although the southern subspecies of the white rhinoceros seems to be secure at the moment, it inhabits a part of the world, South Africa, that is extremely volatile and probably destined for much instability in the future. Two years ago the northern subspecies population was estimated to be about a thousand animals. Now they are down to fifteen. The AAZPA has been working with other conservationists to rescue those fifteen and to establish a captive propagation program, because poaching is rampant in the area where they are located.

The black rhinoceros is another species whose population has declined precipitously, from perhaps 100,000 individuals twenty-five years ago to fewer than 10,000 today.

In contrast, the Indian rhinoceros's population has expanded recently because of effective conservation in the field. This species, however, is located entirely in the wilds of India and Nepal, and the potential instability of the region as well as the tremendous human problems that are occurring there do not provide an optimistic future for this rhino in its natural habitat.

In terms of total number, the Javan rhinoceros might seem to be the most endangered rhino species, but it has not been designated yet for the SSP for several reasons. One is that AAZPA facilities may be able to accommodate MVPs of only four taxa of rhinoceros. The Javan, although rare, resembles the Indian to the extent that the two species are placed in the same genus, whereas all four species designated for the SSP are in separate genera, thus representing a higher level of "uniqueness" than the species. Furthermore, the Javan rhinos seem to be better protected in the

Figure 5.7
A bongo calf born to an eland mother as a result of an embryo transfer. Such cross-species transfers indicate how rapidly this kind of reproductive techonology is advancing. Photograph courtesy of the Cincinnati Zoo.

Figure 5.8
This female Sumatran rhinoceros is the first to be captured as part of a conservation strategy in Malaysia that will include captive breeding.

wild than the Sumatran is. And finally there does not seem to be much prospect of obtaining any founder stock in the near future, especially without detracting from the small but expanding wild population.

And then there is the Sumatran rhinoceros. There are an estimated 300 to 500 of them still in the wild, widely distributed over Southeast Asia. But that is, of course, the problem: They are so widely distributed that their population has become small and fragmented. Much of the population is located outside protected areas. Deforestation continues, poaching is rampant, and Sumatran rhino horn can still be found in the Chinese pharmacies of Singapore. What has been proposed, therefore, is a conservation strategy for the Sumatran rhino that protects natural populations in those few reserves large enough to be viable for long-term survival and that employs the other animals for captive propagation. Five sanctuaries in Southeast Asia might be able to accommodate populations for the long term. Animals outside these areas probably have little or no hope of contributing to the survival of the species.

To salvage the situation, the AAZPA, in conjunction with some British zoos, the IUCN, and the wildlife departments of Malaysia and Indonesia, is proposing a roundup of isolated animals to employ them for captive propagation. The plan is to capture six or twelve pairs of animals, enough for a viable foundation, and to distribute them equally among captive facilities in Southeast Asia, North America, and Europe. If successful, this project might well serve as a model for similar programs with other species. Two females are already in captivity in Melaka on the Malay Peninsula, and another male has been captured in Sumatra.

The zoo ark has thus embarked, but there will be stormy seas ahead. The ark must navigate between two major sources of failure, its own versions of the two evils of Greek legend, Charybdis and Scylla.

The Charybdis for zoos is that financial resources, political will, and public support will not be adequate to keep the ark afloat. Maintaining the zoo ark is too expensive. Bill Conway, the director of the New York Zoological Society and chairman of the SSP Committee, has computed some of the costs that are involved. In 1980, he estimated that maintenance of 2000 species at population sizes of 500 each for twenty years would cost $25 billion, the approximate price tag of the program that placed a man on the moon. These calculations were predicated on an average annual maintenance cost of $625 per animal. Considering food, shelter, artificial temperature control (heat for gorillas; air-conditioning for penguins), and curatorial and veterinary care, I believe that this cost is an underestimate. A more realistic figure is probably $1500 per year per animal. Add to this the research necessary for reproductive technology and the cost for the ark over the next couple of centuries and the total will easily be in the hundreds of billions of dollars, not an intimidating sum perhaps to planners in the Pentagon but quite a concern for zoo conservationists.

To date, zoos have been able to support their conservation efforts largely from existing budgets and resources, but they do it with increasing difficulty. A major problem is that zoos are now assuming global responsibilities in conservation but are still operating with local or parochial constituencies. If there is a decision (and it seems inevitable) by a city council to allocate money to educational programs for children or to conservation efforts for the okapi, the likely answer is clear.

Moreover, it is not just a matter of money. In a coordinated and collective effort to conserve species, there will have to be sacrifice of self-interest. A particular zoo may need to move its solitary but beloved proboscis, or "Jimmy Durante," monkey to another facility for propagation and the good of the species. Or conservation programs may argue that a zoo not invest its resources in a charismatic species such as the gorilla but rather work with primate species that may be less appealing to the public but more in need of help. Convincing local zoo boosters that they should not try for the "gorilla of their dreams" is extremely difficult.

Then, too, there are all the characteristics of the human species that generate personality conflicts and personal ambitions and agendas and that render cooperative versus competitive activity an uneasy enterprise. Conway compares the attempt to organize zoos collectively into the ark as similar to the efforts required to develop the United Nations.

The Scylla for the zoo ark is the danger that the animals will not be quite as "wild or natural" when they emerge from the ark as when they entered. Like Scylla, they may be multiheaded freaks. Or reciprocally, the voyagers might feel as if they were not really on an ark at all but on a spaceship that now has landed on a planet unlike the Earth they knew. The way to prevent changes in the animals is to manage them well, genetically and demographically. But there is little protection against the change of natural habitats and ecosystems. We can try to preserve only as many pieces as possible in the hope of restoring some semblance of nature in the future.

Notes

1. D. Simberloff, "Mass extinction and the destruction of moist tropical forests," *Journal of General Biology* (1985).

2. M. Soule, M. Gilpin, W. Conway, and T. Foose, "The millennium ark: How long a voyage, how many staterooms, how many passengers?" *Zoo Biology* 5(2) (1986), 101–113.

3. United Nations, "Long range global populations projections," *Population Bulletin* 14 (1982).

4. World Bank, *World Development Report* (Washington, D.C.: The World Bank, 1984).

5. G. Hutchinson, *The Ecological Theater and the Evolutionary Play* (New Haven, Connecticut: Yale University Press, 1963).

6. B. Grzimek, *Grzimek's Animal Encyclopedia* (New York: Van Nostrand Reinhold, 1972).

7. J. Raczynski, "Progress in breeding European bison in captivity," in *Breeding Endangered Species in Captivity,* R. D. Martin, ed. (New York: Academic Press, 1977), 253–262.

8. R. Fitter, "Arabian oryx returns to the world," *Oryx* 16(5) (1982), 406–410.

9. M. S. Price, "Reintroduction of Arabian oryx to Oman," in *International Zoo Yearbook* 24 (1986).

10. D. Kleiman, B. Beck, and R. Hoage, "Golden lion tamarins successfully released," *AAZPA Newsletter* 26(4) (1985).

11. R. E. Ricklefs, *Report of the Advisory Panel on the California Condor* (New York: Audubon, 1978).

12. Mark Crawford, "The last days of the wild condor," *Science* 229 (1985) 844–845.

13. T. Foose. "The relevance of captive populations to strategies for conservation of biotic diversity," in *Genetics and Conservation,* C. Schonewald-Cox, S. Chambers, B. MacBryde, and L. Thomas, eds. (Menlo Park, California: Benjamin-Cummings, 1983), 374–401.

14. T. Foose, R. Lande, N. Flesness, G. Rabb, and B. Read, "Propagation plans," *Zoo Biology* 5(2) (1986), 139–146.

15. T. Foose, "Demographic management of endangered species in captivity," *International Zoo Yearbook* 20 (1980), 154–166.

16. T. Foose and U. Seal, "Species survival plans for large cats in North American zoos," in S. Douglas Miller and Daniel D. Everett eds. *Cats of the world: biology, conservation, and management.* National Wildlife Federation.

17. W. Conway, "The practical difficulties and financial implications of endangered species breeding programs," *International Zoo Yearbook* 24 (1986).

Suggested Readings

There are several journals that should be consulted to keep up to date in the area of zoo conservation and captive propagation. They include *Zoobiology, Oryx, Focus (World Wildlife Fund Newsletter),* and the *International Zoo Yearbook.* The AAZPA (American Association of Zoological Parks and Aquariums) conference proceedings are also a good source. Also *Animal Kingdom,* a magazine published by the New York Zoological Society for general readers, contains relevant discussions of endangered animals and zoo conservation. Also, *On the Edge,* a Wildlife Preservation Trust newsletter and the *IUCN* (International Union for the Conservation of Nature and Natural Resources) *Bulletin* are good sources. As for books, the following is a brief list of useful literature.

C. M. Schonewald-Cox, S. M. Chambers, B. MacBryde, and W. L. Thomas, eds., *Genetics and Conservation: A Reference for Managing Wild Animal and Plant Populations.* Menlo Park, California: Benjamin-Cummings, 1983.

M. E. Soule and B. A. Wilcox, eds., *Conservation Biology: An Evolutionary-Ecological Perspective.* Sunderland, Massachusetts: Sinauer Associates, 1980.

U.S. Seal and T. J. Foose, "Development of a master plan for captive propagation of Siberian tigers in North American zoos," *Zoobiology* 2 (1983), 241–244.

O. H. Frankel and M. E. Soule, *Conservation and Evolution.* Cambridge: Cambridge University Press, 1981.

J. F. Crow and M. Kimura, *An Introduction to Population Genetics Theory*. Minneapolis, Minnesota: Burgess Publishing, 1970. This is a good source, but it is technical.

H. Hediger, *Wild Animals in Captivity: An Outline of the Biology of Zoological Gardens*. New York: Dover, 1950.

Appendix

A lot of water has flowed since "Riders of the Last Ark" was originally published in 1986. The flood of destruction for wildlife and wetlands has proceeded inexorably. In 1993, only 2,500 black rhinos survive in Africa, instead of the 10,000 estimated in 1986. Encouragingly, however, captive propagation programs have expanded and improved. We are also building new bulwarks against the flood for threatened populations in the wild and more bridges between captivity and the wild.

In 1986, the captive ark consisted of a few rafts with skeleton crews. Today, there are fleets of programs with growing brigades of increasingly sophisticated scientists and managers to manage captive populations. There have also been advances in solving the two major problems for captive programs: selection of taxa and management of those selected.

In 1986, formal captive propagation programs in various regions of the zoo world were embryonic. The Species Survival Plan (SSP) in North America had designated 37 taxa, for which only a few masterplans had been completed. Regional programs were in even earlier stages of development in the United Kingdom and Australasia, and only just initiated in Europe. In 1993, twelve regions of the zoo world have formally organized programs (figure 5.9). Five of these regions have programs for over 300 taxa and plans for at least 300 more by the end of the decade (table 5.3). While the selection of taxa for captive programs was somewhat ad hoc in 1986, the selection process has now become more systematic through strategic priority processes called the Conservation Assessment and Management Plan (CAMP) and the Global Captive Action Plan (GCAP).

There have also been enormous advances in terms of the second major problem for captive programs, the management of selected taxa. Genetic and demographic management has become more sophisticated. Management is also placing greater emphasis on husbandry as a third component of the captive management triangle (figure 5.10). The importance of husbandry and health is exemplified by the case of the black rhino, although it is not yet totally seaworthy for its ride on the captive ark. This species seems to be accelerating toward extinction in the wild. There are captive programs and reproduction is

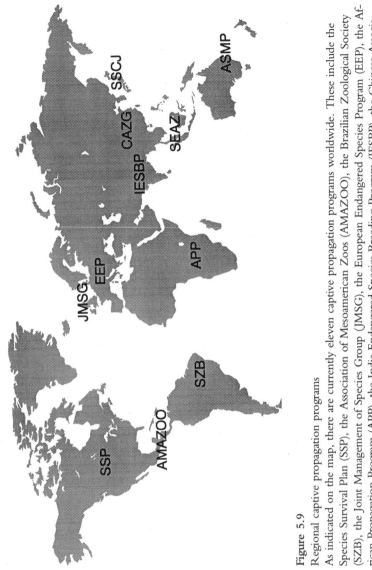

Figure 5.9
Regional captive propagation programs
As indicated on the map, there are currently eleven captive propagation programs worldwide. These include the Species Survival Plan (SSP), the Association of Mesoamerican Zoos (AMAZOO), the Brazilian Zoological Society (SZB), the Joint Management of Species Group (JMSG), the European Endangered Species Program (EEP), the African Propagation Program (APP), the India Endangered Species Breeding Program (IESBP), the Chinese Association of Zoological Gardens and Aquariums (CAZG), the Southeast Asian Zoological Association (SEAZ), the Species Survival Committee of Japan (SSCJ), and the Australasian Species Management Program (ASMP).

Table 5.3.

Regional captive propagation programs worldwide, 1992

Region	Designated and pending taxa					
	Mammals	Birds	Herps	Fish	Invrt	Total
Australasia Species Management Program (ASMP)	23	11	3	0	0	37
European Endangered Species Program (EEP)	67	16	2	0	0	85
Joint Management of Species Group (JMSG)	65	27+	51	0	25	168+
Species Survival Plan (SSP)	52	14	6	29	9	110
Species Survival Committee of Japan (SSCJ)	27	12	1	11	0	51

Note: The table includes overlap between the different regions, i.e., the same taxa can be in one or more regions. With this in mind, the total number of unique mammal taxa in propagation programs worldwide is mammals: 13, birds: 60 or more, herps: 56, fish: 40 and invertebrates: 27, with a grand total for the number of unique taxa in these programs of over 314. Another way of looking at this is that there are 230 taxa unique in only one region, 49 taxa unique in two regions, 17 taxa unique in three regions, 13 taxa unique in four regions and 5 taxa unique in five regions.

India has also initiated movement toward regional captive propagation programs by establishing studbooks for 44 mammal, 15 bird, and several herp taxa.

reliable, but mortality is unsatisfactorily high due to a curious syndrome characterized by hemolytic anemia (disintegrating red blood cells) and skin ulcerations. To solve the problem, the zoo community has been collaborating with researchers in human medicine who have discovered that the syndrome resembles a condition caused by enzyme deficiencies. A cure may be in sight.

Just as significant as the advance of the science is the increase in the number of persons trained in the science and therefore capable of formulating propagation masterplans. As a result, there has also been a tremendous increase in the number of masterplans providing scientific recommendations on an animal-by-animal and institution-by-institution basis for management and propagation. This training has become institutionalized through such developments as the American Association of Zoological Parks and Aquariums (AAZPA), Conservation Academy, and Small Population Management Advisory Group

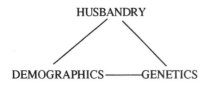

Figure 5.10
Captive management triangle.

(SPMAG) in North America, with similar activities developing elsewhere in the zoo world.

Reproductive technology and genome banks have also progressed. Since 1986, artificial insemination and in vitro fertilization have succeeded with the tiger. In fact, the encouraging advance in this technology is one of the reasons for reduction in the length of time used to establish target Minimum Viable Populations (MVPs) for captive programs. One hundred years is used as the standard objective, instead of two hundred years. Despite the successes, challenges remain. Routine use of reproductive technology is still several years in the future for the rhinos, a group that desperately needs this tool for their conservation.

The 1986 version of "Riders of the Lost Ark" discusses ebbs and flows in the fortunes of species in the wild. The proximate or distant cause of these ebbs and flows is always human activity. A prime case in point is the Siberian tiger, a model for development of captive programs and hence a focus of the 1986 paper. The changes that have occurred in Russia have had a dramatic effect on the fortunes of the Siberian tiger in the wild. Ironically, under the authoritarian regime of the Soviet Union, the tiger received excellent protection. The tiger populations prospered to the point where they exceeded the capacity of their reserves, although total numbers remained about 300. Now, with the reduction of regulations and with the economic difficulties occurring in Russia, conflicts with humans have increased. Tigers are being excessively exploited for their bones and other parts, which are sold for high prices in the Far East. Captive propagation of this subspecies becomes increasingly more important.

More than ever, small wild populations like the Siberian tiger require intensive management similar to that which is applied in captivity. Another significant development over the last seven years has been the evolution of Population and Habitat Viability Analyses (PHVAs) to assist in identifying the need, and formulating the plans, for such intensive management of small populations in the wild. Members of the zoo community have developed PHVAs to a great extent, making them a bridge between captivity and the wild.

A PHVA is an intensive analysis of a particular taxon or one of its populations. PHVAs use computer models for several purposes: to explore extinction processes that operate on small and often fragmented populations of threatened taxa; to evaluate a range of scenarios for the populations under a variety of management (or nonmanagement) regimes that are proposed in the PHVA process; and, as a result, to recommend management actions that maximize the probability of survival or recovery of the population. The management actions may include: establishment, enlargement, or more management of protected areas; poaching control; reintroduction or translocation; sustainable use programs; and captive breeding.

Rescue and restoration of a species exterminated in the wild is another kind of bridge, and there has been a spectacular case since 1986. The black-footed ferret, once ubiquitous on the Great Plains of North America, has had a fitful history over the last century. Feared extinct in the late 1970s, a single small population was rediscovered in the early 1980s. As a single small population, the remnant of this species was at high risk of extinction because of loss of genetic diversity and vulnerability to environmental threats. At first, attempts to achieve recovery of this population concentrated on protection in the wild and deferred development of a captive program. However, after some initial increase, the population was afflicted by two disease epidemics, one decimating the ferrets themselves, the other devastating their prey, the prairie dog. About the time *The Last Extinction* was published, the last known 18 ferrets were moved into captivity. By 1991, the captive population had increased to about 300 animals and reintroduction to the wild was commenced. While it is too early to declare that the species has been saved, its extinction has certainly been prevented for now, and progress toward restoration in the wild continues.

Just as there are ebbs and flows for species in the wild, there are ups and downs for species in captivity. The 1986 chapter concluded with an optimistic preview of the newly developing captive program for Sumatran rhinos. At that time, only 3 rhinos had been rescued. Since then, 31 more rhino have been captured in Malaysia and Indonesia for the captive programs in those countries as well as in the United States and the United Kingdom. Unfortunately, 12 of these rhino have died. One calf has been produced, and although most of the gestation occurred in captivity, she was conceived in the wild. There are currently 24 animals (9 males and 15 females) in captivity in four countries. The unfulfilling performance of this program is due to a number of factors, including difficulties in capturing mature males and insufficiently vigorous management to permit and promote reproduction. The challenge continues.

Finally, it should be observed that while captive propagation is ever growing in importance, these programs are more and more being developed as part of more holistic strategies for species conservation. Through technology transfer as well as financial support of in situ conservation, the captive community is not only providing an ark but also trying to ensure that the ark has places to land.

Additional Readings

Many additional publications for further reading have been developed over the last seven years. Among them are the journal *Conservation Biology,* the newsletter *Captive Breeding Specialist Group* (CBSG) *News,* and the book *Last Animals at the Zoo* by Colin Tudge, Island Press, Washington, D.C. The large and growing number of Conservation Assessment and Management Plan (CAMP), Population and Habitat Viability Analysis (PHVA), and Global Captive Action Plan (GCAP) reports can be obtained from the CBSG Office, 12101 Johnny Cake Road, Apple Valley, MN, 55124, USA. A copy of a current list can be obtained from CBSG.

Finally, as an update on the 1986 chapter suggested readings, *Animal Kingdom* magazine is now called *Wildlife Conservation.*

Sharing the Earth with Whales

Norman Myers

*Genius in the sperm whale? Has the sperm whale ever written a book, spoken a speech?
No, his great genius is declared in having done nothing to prove it.*

Herman Melville, 1851

*The sperm whale brain has inexhaustible possibilities for establishing links between stim-
uli and forms of reacting, also the accumulation of individual forms of reaction which are
volitional or conscious, elaborated and accumulated by the animal during its life. The
sperm whale can be said to be a "thinking" animal, capable of displaying high intellectual
abilities.*

A. A. Berzan, Soviet biologist, 1971

Whales take up no space on Earth that is wanted by humans. So they are
more fortunate, in that respect at least, than are elephants. Nor do they
consume vegetation that humans would prefer to see going down the
throats of cattle. In that respect, too, they are more fortunate than zebras.
Nor do they kill humans' livestock, or indeed humans themselves. So they
are more fortunate than tigers. Wretched whales, however, that their bodies
contain materials sought by humans. Through this quirk of physiology, they
have been brought to a parlous state, with numbers only a small fraction
of what they once were.

   Much has been written, and 'twere bootless to repeat it here, about
the reasons, and excuses, for the whales' present predicament. Suffice it to
say that it has less to do with the feckless greed of individual humans and
more to do with the institutional myopia of humans collectively. The
intrinsic problem lies, in my view anyway, with the "commons" nature of
the whales' environment, with marketplace mechanisms, with discount
rates, and with sundry other such boring factors that are not amenable to
attack from small boats (much as the Greenpeacers are admirable beyond
description) (Clark 1977; Myers 1975). This sort of institutional problem
could be resolved at a stroke, as has been recently suggested, by declaring

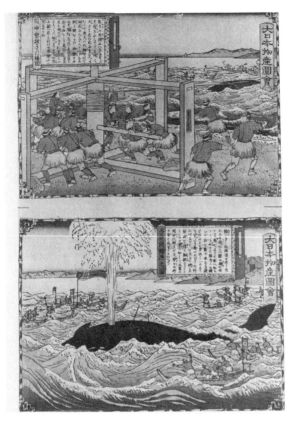

Figure 6.1
"Catching Whales in Iki Province" by Hiroshige III (Hiroshige Shigemasa) (1834–
1894). Courtesy of The Kendall Whaling Museum, Sharon, Massachusetts.

all whales the formal property of humankind as of a certain date, let's say
January 1, 1994, with property rights vested in a United Nations body—
whereafter anyone wishing to exploit a whale, whether consumptively or
not, would have to purchase an exploitation permit at the latest market
price. In turn, the market price would be determined by all consumers in
the marketplace, including conservationists who wish to out-bid Japanese
whalers for a whale permit and then sit on it (or ceremoniously burn it in
the global village square).

This chapter supposes, then, that the rational management of whales
is not in question. In the main we know what to do. We simply do not
get on with it, which makes it another matter. What is of acute concern,
albeit little recognized, is that whales may yet have to face a greater threat
than any to date. It lies with the gross degradation of their environments

and destruction of their life-support systems. Therein they will learn to suffer alongside the elephants and zebras. It is the thesis of this chapter that given the way we run our planetary affairs and exploit our biosphere, we are effectively saying there is no room on Earth for both ourselves and whales. And "we" means all of us, not just the Japanese.

Fortunately, there is an upbeat side to the situation as well. There is just time for us to turn the corner and determine to safeguard our biosphere as if it is the only one available to us, not as if we have a spare biosphere parked out in space to which we can move if ever we find we have over-done things on Planet Earth. For sure, we shall not set our biospheric house in order with the whales in mind. But we shall thereby save the whales and allow them to continue to share the One-Earth home with us.

The Present Status of Whales

First let us remind ourselves of the parlous state of whales. According to mid-1989 estimates (Holt 1989; see also Bonner 1989; Cousteau and Paccalet 1989; Wursig 1989), based on a review of eight years of annual sighting surveys, the situation is even worse than feared. In one of the whales' principal habitats, the southern hemisphere from south of 60 degrees S. to the edge of the Antarctic ice, the blue whale was estimated in the 1960s, at the time of its first full protection, to have declined from an original total of almost 250,000 individuals to somewhere between 20,000 and 10,000 individuals. But the latest findings suggest the true figure today is probably 1,100 at most (with an absolute maximum of 2,000) and possibly as low as 200 individuals.

Fin whales were protected in the 1970s, when it was believed their original number of about 500,000 had declined to only 100,000. But the new estimate, based on more systematic surveys, suggests only about 2,000 remaining, within a range of 1,600 to 2,400 (and an absolute maximum of 4,000). A similarly low figure is now accepted for humpback whales. Sei and Bryde's whales are only a fraction of what they once were. As for sperm whales, as recently as March 1989, the International Whaling Commission was still asserting that of an original stock of 1.25 million individuals in the southern hemisphere, some 950,000 remained. But the latest analyses suggest only 400,000 sperm whales in all the oceans, possibly far fewer. While this might sound like a large number, it must be considered a serious setback as compared with what was once there.

Moreover, these figures raise critical questions about whether populations reduced to sparse numbers can ever recover. After all, their thin spread over extensive habitats means there could be difficulties in such basic

Figure 6.2
American sperm-whaling. "Taking a Whale/Shooting a Whale with a Shoulder
Gun" by Robert Walter Weir, Jr. (1836–1905). Courtesy of The Kendall Whaling
Museum, Sharon, Massachusetts.

factors as breeding relationships. Can enough individuals of either sex find
each other for mating purposes? The humpbacks have been protected for
years, yet there is no evidence that they are recovering. The same applies
to the bowhead and blue whales. Fortunately the gray whale seems to be
showing a slow increase in numbers.

To reiterate a key factor, these estimates of species numbers are rough-
and-ready in the extreme. Whales are difficult to sight, and they are spread
out over vast stretches of the oceans. It is more than challenging to come
up with even approximate estimates of their present conservation status and
their future survival outlook. In face of much uncertainty, however, let us
remember that it is the prudent course to err on the cautious side. After
all, the early estimates of ozone-layer depletion have been repeatedly found
to be on the low side.

The Marine Realm

While population estimates for many whale species are still subject to much
uncertainty, we are even more uncertain about the nature of their marine

environment. Recall the story of two people who were looking out for the first time on the Pacific Ocean. The first, struck by the vastness of the expanse, cried "Look at all that water!" The second responded, "Yes, but remember that's only the surface." Let's bear in mind, moreover, that we now know more about outer space than about life in the deep ocean.

What we know for sure is that the marine realm features a wide array of life forms. True, there are relatively few species. If we use a minimum estimate for all species on Earth, 10 million, the seas are believed to contain only about one-fifth of the total. But this is a restricted way of looking at the situation (Ray 1985). We should also consider the comparative diversities of different taxa as between the terrestrial and marine realms. Not only are all phyla of life present in the seas but the majority of classes of animals are exclusively marine. At these two taxonomic levels, the marine realm reveals perhaps twenty times more biodiversity than there is on land.

Yet we still perceive the seas with a land-based outlook. Our research strategies for ecological analysis are largely driven by terrestrially derived models (Longhurst and Pauly 1987). This sizable lacunae notwithstanding, the seas deserve to be understood in their own right: after all, we should call our globe Planet Ocean. But we are still a long way from a united scientific perspective of "oceanology," a challenge that is all the more urgent if we are to accommodate for problems of human ecology during the course of marine research (Ray 1985). To be technical a moment, the challenge

Diversity

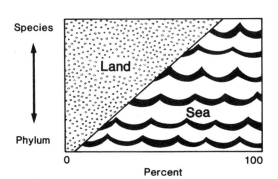

Figure 6.3
Comparison between diversity of land and sea at different taxonomic levels. Source: "Man and the Sea," G. Carleton Ray. *American Zoologist* 1985, vol. 25 p. 455.

is to develop unifying, ecosystem-based approaches to oceanology—as for our uses of the oceans and their resources. Without a proper scientific understanding of marine processes, we shall not be able to do a proper job of conservation management.

Yet we have only a rudimentary grasp of one of the key functions of the planetary ecosystem: the ocean/atmosphere interface (Broecker 1987; Tyack 1989). We are far from knowing, let alone understanding, the role of the ocean in modulating global change in the atmosphere and climate. Most of our models for predicting climate change take the oceans into scant account, mainly because we do not know enough about the complex biological, physical, and chemical processes involved. Even when we consider the oceans apart form the rest of the planetary ecosystem, our lack of precise knowledge continues to surprise us. Conventional data sources, such as ships and buoys, showed that for the period 1982–88 the global oceans were warming up by about 0.5°C per year. But recent measurements have been taken from more comprehensive and systematized sources of information that include infrared sensors aboard National Oceanic and Atmospheric Administration (NOAA) satellites reading from 2.5 to 3 million points in the ocean, by contrast with only 50,000 conventional thermometer readings from ships and buoys. These recent readings reveal that the rate of temperature rise during the 1980s has been about twice the earlier estimates (Strong 1989).

Moreover, the oceans serve as a vast sink for atmospheric carbon dioxide. Of all anthropomorphic carbon released into the atmosphere each year, some eight billion tonnes (a tonne, or metric ton, is approximately equal to 1.102 U.S. tons) (Schneider 1989), the oceans are thought to be absorbing about one half. But as the atmosphere grows warmer in the wake of the greenhouse effect, the oceans may start to release more carbon into the atmosphere than they take out of it. At higher latitudes in particular, the oceans absorb much carbon dioxide by virtue of their colder waters, and the North Atlantic is a key region where the atmosphere and the ocean exchange exceptionally large quantities of carbon dioxide. Yet the highest increase in sea-surface temperature is likely to be in the North Atlantic.

Ocean Ecosystems: Pollution and Habitat Destruction

A prime source of concern for marine mammals lies with waterborne contaminants. In 1987 hundreds of dolphins died along the eastern seaboard of the United States, ostensibly because they had been weakened by pollutants, whereupon they became all the more susceptible to viruses and bacteria (Joyce 1989). In the same year a number of beluga or white whales died in the St. Lawrence estuary, their bodies revealing high concentrations of DDT, PCBs, mercury, cadmium, carcinogenic chemicals, and other toxins, causing most of the whales ultimately to die of septicemia. Farther southward many humpback whales died too, because they had been eating mackerel containing toxic algae. On the opposite side of the Atlantic in the Barents Sea, there was a decline in biological productivity overall (especially with respect to cod fisheries), a phenomenon described by Norwegian and Russian scientists as "an ecological disaster."

In 1988 dolphins and seals died by the thousands in the North Sea, the Irish Sea, and the Baltic Sea, apparently because of a virus. Algae blooms off the coasts of Sweden, Denmark, and Norway killed virtually all marine life in their paths (including at least $3.5 million worth of farmed salmon and trout). There was a plague of algae in the Adriatic Sea as well, reportedly the worst ever, attributed to phosphate eutrification from the River Po.

A series of other catastrophic incidents have affected marine life, notably in the North Atlantic and the adjoining seas of the Mediterranean, Baltic, and Barents, with each phenomenon being described as "unprecedented" (Evans 1987). It is realistic to suppose that increasing amounts of pollutants will spread in marine ecosystems.

Marine pollution is especially a problem in bays and lagoons, which also suffer from "development." Gray, humpback, and right whales all use shallow coastal lagoons and bays for breeding. Such localities are often restricted in extent, so sudden damage to these habitats can prove serious. The San Diego Bay, once a major breeding ground for gray whales, no longer serves that function, since intense human activity has driven them away. Inlets of the Baltic Sea are now so polluted that whales, abundant there at the turn of the century, have completely disappeared.

Also vulnerable are feeding areas for coastal whale species. For instance, hydroelectric developments on the Manicouagan River in eastern Canada are thought to have seriously affected white whale populations in the outer St. Lawrence estuary by closing off important feeding grounds. In Arctic waters, there could be sizable threats from offshore oil exploitation, affecting particularly white whales, narwhals, and bowheads (Doak 1988).

Serious and harmful as these examples of pollution and habitat depletion are, they are surely a small portent of what lies ahead. Coastal zones are highly favored by human communities, with often deleterious effects for offshore marine communities. Consider, for instance, the Caribbean

Sea, where one-sixth of the world's oil is produced or shipped. Supertankers and offshore oil rigs inject more than 100 million barrels of oil into the sea each year.

Because we have not yet developed a comprehensive and integrated approach to management of the land/sea interface, we tend to use the marine realm as a dumping ground for all manner of toxic materials. We should surely expect there will be more biotic crises, growing in intensity and geographic spread.

While many of the above incidents can be attributed, partially at least, to heavy metals, toxic chemicals, and other lethal pollutants, there could be a further disruptive factor for marine biotas in the warming that is already overtaking the oceans. Several of the last few years have seen unusually warm ocean weather in the northeast Atlantic and the Mediterranean Sea—precisely the sort of weather that, in conjunction with excess phosphates and nitrates, can trigger algal blooms. Mild weather also serves as a known source of stress for the thermoregulatory system of seals, and probably of other marine mammals, too.

This final factor of warmer temperatures brings us to the greatest potential disruption of all for marine ecosystems, the greenhouse effect.

The Greenhouse Effect and Global Warming

As noted, the oceans are warming up twice as fast as had been believed. Obviously water temperature has a profound influence on food stocks. Colder sectors of the oceans, notably areas of upwellings where currents meet, are exceptionally productive; and these areas are utilized by many cetacean species for seasonal, if not year-round feeding. Resident species, particularly those with coastal habitats like the four *Cephalorhynchus* species (dolphins) in the southern hemisphere, could find themselves left with no adjacent productive areas into which they could migrate if their present localities no longer produce food of the right amounts and sorts (Klinowska 1989a).

Of course one might suppose that many cetaceans, being accustomed to roaming widely around the oceans, could readily migrate in response to climate change such as global warming. But the situation may not be so straightforward. What if the environmental circumstances to which marine communities are adapted cannot be replicated elsewhere? Consider the case of krill, which plays a key role in the food chain of the greatest aggregation of whales in the world, located in the southern ocean. The extraordinary productivity of krill, around 250 million tonnes per year (Evans 1987),

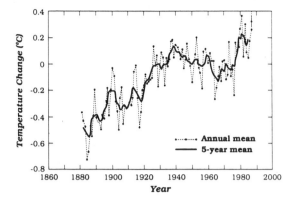

Figure 6.4
Variations in average temperature of the globe (land and sea) expressed as departures from the mean over the last thirty years. Source: "The Message from the Oceans," John H. Steele. *Oceanus* 1989 vol. 32:2 p. 6.

reflects a number of localized conditions, notably the convergence of warm and cold currents in the southern ocean, leading to a nutrient-rich upwelling around the shores of Antarctica. Under the impact of the greenhouse effect, the site of the currents' convergence would want (so to speak) to move southward. It would be unable to do so because of the Antarctic land mass in the way. How far the upwelling would continue anyway, with a much warmer current from the north washing up against a somewhat warmer current in the south, is difficult to say. But the upwelling, with all it means to krill productivity, could be significantly disrupted if not displaced.

Ecological Discontinuities in Ocean Ecosystems

Apart from temperature increase, a number of other physical changes predicted for the oceans have the potential to drastically influence marine mammals, indeed all biotas. Such physical changes include shifts in salinity, pH, turbulence, upwellings, storms, sea ice, and global circulation patterns. Moreover, we should not expect that multiple and macrolevel changes will overtake the oceans in linear, readily predictable fashion. Rather, the record shows that the climate change generally, and oceanic reactions in particular, tend to occur in irregular fashion, often through a series of leaps (Broecker 1987).

These "jump effects" and their threshold responses, with potential breakpoints of irreversible injury to marine ecosystems, occur when environmental conditions change in a manner, or on a scale and with a speed, to which ecosystems are not adapted. They may well have been absorbing ecological stresses over a long period without much outward sign of damage—but then they reach a disruption level at which a jump event becomes increasingly likely and ultimately inevitable (in terrestrial ecosystems we have seen how this can happen with acid rain).

Alternatively stated, the stresses build up covertly over a number of years, before suddenly revealing themselves with critical impact. Moreover, when the stresses are removed, the ecosystem may not return to its former equilibrium state, no matter how much we may try to restore the injury. Instead it may well settle into a new equilibrium, one that may be less useful to our purposes, much less to whales' survival—that we may not know how to amend, no matter how much we might be ready to pay to make good the damage.

Another pronounced change for ocean ecosystems could stem from depletion of the ozone layer. The thinning layer allows increased amounts of ultraviolet radiation (UV-B) to reach Earth's suface; and few groups of organisms are more susceptible to UV-B injury than phytoplankton in the seas' surface layers (other marine species vulnerable to UV-B injury are copepods among other primary consumers, plus the larvae of several fin and shellfish species). Were phytoplankton to decline in quantities and distribution, there would probably be two pronounced repercussions. First there would be a ripple effect throughout many marine food chains—with all that implies for the larger creatures at higher trophic levels, notably cetaceans. The second consequence would lie with phytoplankton's role in the global carbon cycle. As noted, the oceans are thought to absorb about half of all carbon emitted into the global atmosphere each year, a vital role being played by the photosynthetic activities of phytoplankton. Were phytoplankton communities to be unduly depleted, their carbon-takeup activities would in turn be reduced. In turn again, this could lead to less carbon being absorbed at the oceans/atmosphere interface, with all that could mean for the scope and onset of the greenhouse effect—with, in turn yet again, all manner of further repercussions for marine ecosystems generally.

All this would be imposed, moreover, on whale ecosystems that have already been grossly disturbed. A salient instance lies with food chains in the Southern Ocean, based as they are on krill. The feeding grounds of the great baleen whales in the southern oceans are more productive than the finest agricultural lands of the United States: a single hectare can sustainably produce more than one tonne of animal protein per year in the form of krill and zooplankton. Krill productivity is around 250 million tonnes per year; to put the figure in proportion, it is two and a half times as much as the annual yield of the world's fisheries, some 100 million tonnes.

After the great whales were drastically reduced in numbers, other krill-eaters multiplied and moved in to fill the ecological space. Today krill is consumed by an increased abundance of smaller whales (minkes), seals (notably the crabeater seal), and seabirds (notably several penguin species)— to say nothing of the swiftly developing krill fisheries. In the year 1900 the great whales were probably consuming 190 million tonnes of krill per year, but today they are estimated to account for only 40 million tonnes. Meanwhile seals have increased their numbers, expanding their consumption of krill from 50 to 130 million tonnes per year, and seabirds consume roughly the same.

So if the great whales survive, the ocean ecosystems they will inherit will no longer be the same as those of yesteryear. There has been a profound and pervasive shift in the whole web of life in the southern oceans.

The Mediterranean Action Plan

Amidst a welter of adverse reflections, let us remind ourselves of a piece of good news, to demonstrate that we can get our marine act together even in unpromising circumstances. Consider the Mediterranean Sea, which encompasses only 1 percent of the globe's ocean surface yet features 50 percent of all marine pollution. The pollution has been threatening a $10 billion-a-year tourist industry, as well as a number of sizable fisheries.

In the mid-1970s the United Nations Environment Programme convened a conference, with all eighteen coastal nations except Albania participating. The upshot has been a joint response to a joint problem: through widespread cleanup measures, the Mediterranean is now starting to recover. The common effort has been all the more remarkable in that the participant parties have included such traditional foes as Greece and Turkey, Israel and Syria, Egypt and Libya, France and Algeria, and Spain and Morocco. Indeed the cleanup plan has been so successful that it is now being used as a model for ten other "regional seas" initiatives (Haas 1990; Grenoin and Batisse 1991).

What's at Stake

As this chapter demonstrates in dozens of doleful ways, the prospect for whales and other cetaceans is distinctly unpromising. Yet a world without whales is surely unthinkable. They represent some of the most advanced

Know Your Whales: Their Names,

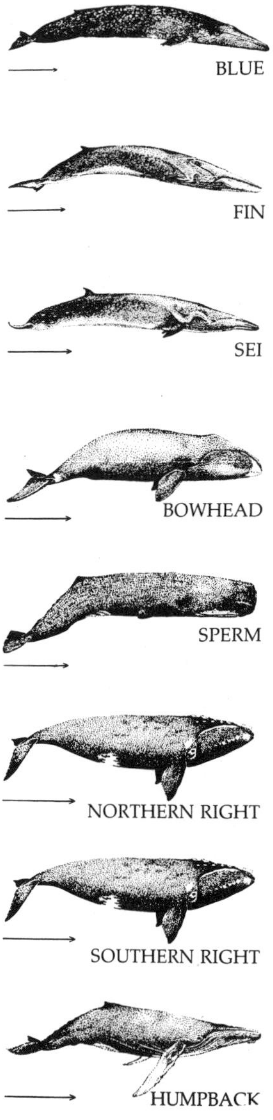

Common Name	Scientific Name	Derivation
		All derivations from Latin, except those marked (Gk) = Greek, and (ME) = Middle English.
Blue	Balaenoptera musculus	balaena = whale, pteron = wing or fin, mus = mouse[a]
Fin	Balaenoptera physalus	physalos (Gk) = rorqual whale
Sei	Balaenoptera borealis	boreal = northern
Bowhead	Balaena mysticetus	mystakous = moustache, cetus = whale
Sperm	Physeter catodon, or P. macrocephalus	physeter (Gk) = blower, kata (Gk) = inferior, odontos (Gk) = tooth, makros (Gk) = long, kephale (Gk) = head
Northern right	Eubalaena glacialis	eu = right or true, glacialis = icy or frozen
Southern right	Eubalaena australis	australis = southern
Humpback	Megaptera novaeangliae	megas = large, novus = new, angliae (ME) = English
Gray	Eschrichtius robustus	Eschricht = a 19th-century zoologist, robustus = oaken or strong
Bryde's	Balaenoptera edeni	Eden = a 19th-century British Commander
Minke	Balaenoptera acutorostrata	acutus = sharp or pointed, rostrum = beak or snout
Killer	Orcinus orca	orcynus = a kind of tuna, orca = a kind of whale
Pygmy right	Caperea marginata	caperea = to wrinkle, marginata = to enclose with a border
Narwhal	Monodon monoceros	monos = one or single, oden = tooth, keros (Gk) = horn
Beluga	Delphinapterus leucas	delphinos (Gk) = dolphin, a- = without, pteron = fin, leukos (Gk) = white

BLUE

FIN

SEI

BOWHEAD

SPERM

NORTHERN RIGHT

SOUTHERN RIGHT

HUMPBACK

[a]Probably named in jest. Musculus is actually the diminutive form of mouse.
[b]Including a well-established estimate of 7,000 in the northern hemisphere.
[c]Estimate is well established.

Illustrations by E. Paul Oberlander

Figure 6.5

Source: *Oceanus* 1989 vol. 32:1 pp. 12–13. This chart includes amendments to the chart originally published in the Spring 1989 issue of *Oceanus Magazine*, (volume 32, number 1), Woods Hole's international magazine of marine science and policy.

It is very difficult to give even "dependably rough" estimates for whale populations today, given the problems of taking a census of populations in the open oceans and the use of mathematical models. The general sources of the estimates tend to generate figures that are on the high side.

Population Estimates, and Status

Population Estimate		Status Listing		
Pre-exploitation	**Present**	**United States Government**	**International Union for the Conservation of Nature**	
All estimates are from the International Whaling Commission, and except those noted, are highly speculative.				GRAY
228,000	14,000	Endangered	Endangered	
548,000	120,000	Endangered	Vulnerable	BRYDE'S
256,000	54,000	Endangered	Not listed	
30,000	7,800	Endangered	Endangered	
2,400,000	1,950,000	Endangered	Not listed	MINKE
No estimate	1,000	Endangered	Endangered	
100,000	3,000	Endangered	Vulnerable	
115,000	10,000[b]	Endangered	Endangered	
More than 20,000	21,000[c]	Endangered	Not listed	KILLER
100,000	90,000	Not listed	Not listed	
140,000	725,000[d]	Not listed	Not listed	PYGMY RIGHT
No estimate	No estimate	Not listed	Not listed	
No estimate	No estimate	Not listed	Not listed	NARWHAL
No estimate	35,000	Not listed	Insufficiently known	
No estimate	50,000	Not listed	Insufficiently known	BELUGA

[d]Including well-established estimate of 600,000 in southern hemisphere.

The harpoon in each drawing represents five meters.

Although the estimate for blue whales listed in the chart is 14,000, the current level could have declined below 2,000 and possibly as low as 1,100 individuals. The finback, listed at 120,000, could conceivably have declined to a remaining few thousand, while the true present figure for the sperm whale could be far below the 1,950,000 listed here.

The northern right whale (*Eubalaena glacialis*) is now estimated at 350, not the 1,000 whales listed in the original version of the chart. Humpback population estimates are now placed closer to 20,000 than the 10,000 originally cited.

creatures ever to grace the face of the Earth. They are demonstrably intelligent—so far as we understand them. But having evolved in an environment entirely different from our own, it would be reasonably intelligent of us to suppose that they are intelligent in a manner different from our own. And having flourished for millions of years before the advent of the one species able to drive them to extinction, their survival rests in the hands of the one species with the capacity to ensure their continued flourishing in the world's oceans.

As has been often surmised (Klinowska 1989b; Scarff 1980; Scheffer 1989), whales may even possess something we could term a "culture." There is emergent evidence that they have built up a form of "collective wisdom" that is passed on from one generation to the next by older members of herds—which means that by killing the larger and older individuals, whalers could be causing whole population structures to be disrupted, and their rate of recovery to be set back, perhaps terminally. Certain whale species appear to possess a biological sonar or echolocation system that they presumably use to explore their surroundings. This means they could even be so highly developed that they can use their exploratory capacities to "see inside" each other; and if this is true, one whale could tell whether a companion is tense or relaxed just by sensing its heartbeat. So while we have been testing certain individual whales to see if they can pick up our communications, it is just possible they have been communicating all the time with each other in ways that we can barely comprehend.

Yet even as we behold the whales with wonder, and with wondrous ignorance, we are reducing their survival prospects in two salient ways. First, we have directly depleted their numbers through deliberate overhunting—a sort of piracy of the modern age. Second, we have been degrading their environments and disrupting their life-support systems through unwitting forms of destruction. It could eventually be the case that we visit more injury upon whales through the aerosol can than through the harpoon.

All this says much about ourselves. We are apparently prepared to eliminate one of the finest adornments of our planet, and do it casually. If a zoologist were to plumb the depths of some oceanic trench and find there a bottom-dwelling creature weighing 50 tons, the discovery would make the television news and newspaper headlines. Of course, a creature that confines itself to the ocean floor (and thus remained undiscovered) would, almost by definition, be a far less advanced creature than a whale, with limited sensory organs to respond to light, warmth, and the other stimuli associated with evolution's more sophisticated products. It might well show none of the social organization, the ecological relationships, and the other evolutionary refinements that characterize the whales. All the same, it would instantly become a cause célèbre, and not just with scientists and conservationists but with the public at large as well. Yet the last of the great whales are being allowed to slip over the edge—nay, are being shoved over the edge—with hardly a headline. We are ostensibly saying that yes, we would like to find whales some remaining space in our One-Earth home, but no, there's just not enough room right now, and anyway we're too busy with inflation and trade problems.

Or are we starting to say something else? The October 1988 incident of three gray whales trapped in the Arctic ice seemed to resonate with a public sense that things are far from right with whales. The outpouring of support on the part of citizenry around the world, plus the $800,000 spent by government bodies and conservation agencies, seemed to tell a different story. True, the scientist can protest that the public and governments alike seem to confuse the saving of a few individuals with the saving of a species. All the same, the incident stirred the sentiments of millions of people in lands everywhere: surely the individual whales, beleaguered as they were, served as symbols of whaledom, beleaguered as it is?

It is this role of whales, to stand as a measure of our global conscience, that should give us hope. Since whales occupy oceans the world over, they are universal standards by which we can assess our stewardship of the planet. Perhaps whales have still to serve their most significant function for us (human-centered concern as that is!) by reflecting the greatest challenge of our time: to keep the planet in tolerable working order. So can we still share the Earth and its dwindling resources—including some of the most critical life-support systems—with whales? And if it turns out we implicitly decide against the whales, who will vote thumbs up for us? By standing by whales as "global citizens," we may learn what it is to be equal lords of all creation.

John Donne, are you getting a hearing—at long glorious last?

References

Bonner, N. 1989. *Whales of the World*. Blandford Press, London, U.K.

Broecker, W. 1987. Unpleasant Surprises in the Greenhouse? *Nature* 328: 123–126.

Clark, C.W. 1977. *Mathematical Bioeconomics: The Optimal Management of Renewable Resources*. John Wiley, New York.

Cousteau, J. and Y. Paccalet. 1989. *Whales*. Harry N. Abrams Inc., New York.

Doak, W. 1988. *Encounters with Whales and Dolphins.* Hodder and Stoughton, London, U.K.

Evans, P.G.H. 1987. *The Natural History of Whales and Dolphins.* Christopher Helm, London, U.K.

Grenion, M. and M. Batisse. 1991 *The Mediterranean Basin Blues Plan.* Oxford University Press, New York.

Haas, P.M. 1990. *Saving the Mediterranean: The Politics of International Environmental Cooperation.* Columbia University Press, Irvington, New York.

Holt, S. 1989. Even Pessimists Rub Eyes Over New Whale Figures. *BBC Wildlife Magazine* (JULY): 473

Joyce, C. 1989. Poisonous Algae Killed the Atlantic Dolphins. *New Scientist* (Feb.11): 31.

Klinowska, M. 1989a. *Whales, Dolphins, and Porpoises—Survival and Distribution in a Warmer World.* Research Group in Mammalian Ecology and Reproduction, University of Cambridge, U.K.

Klinowska, M. 1989b. How Brainy Are Cetaceans? *Oceanus* 32: 19–20.

Longhurst, A.R. and D. Pauly. 1987. *Ecology of Tropical Oceans.* Academic Press, New York.

Melville, H. 1851. *Moby Dick.* Harper and Brothers, New York.

Myers, N. 1975. The Whaling Controversy. *American Scientist* 63: 448–455.

Ray, G.C. 1985. Man and the Sea—the Ecological Challenge. *American Zoologist* 25: 451–468.

Rodriguez, A. 1981. Marine and Coastal Environmental Stress in the Wider Caribbean Region. *Ambo* 10: 283–294.

Scarff, J.E. 1980. Ethical Issues in Whale and Small Cetacean Management. *Environmental Ethics*: 241–279.

Scheffer, V.B. 1989. How Much is a Whale's Life Worth Anyway? *Oceanus* 32: 109–111.

Schneider, S.H. 1989. The Greenhouse Effect: Science and Policy. *Science* 243: 771–781.

Strong, A.E. 1989. Greater Global Warming Revealed by Satellite-Derived Sea-Surface-Temperature Trends. *Nature* 338: 642–645.

Tyack, P.L. 1989. Let's Have Less Public Relations and More Ecology. *Oceanus* 32: 103–108.

Wursig, B. 1989. Cetaceans. *Science* 244: 1550–1557.

Life in the Next Millennium: Who Will Be Left in the Earth's Community?

David Ehrenfeld

The future is shy. If you want to catch a glimpse of it, you have to sneak up on it from behind. So the place to start for a look into the future is the past. The question of which species of living organisms will be left in Earth's communities in the years to come is not a fit subject for casual futurology. We have to lay a good foundation, and perhaps the best way to begin is with the relatively recent idea of changing the world.

Indeed the idea that we humans can change profoundly the physical and biological world is not old; many people who consider themselves altogether modern still have not grasped it. That the Earth can change, most but not all of us accept: we see the work of earthquake, volcano, drought, and flood. Most cultures have a story of a great flood that wiped out a vast number of plants, animals, and people. The nineteenth-century naturalist and mining engineer Thomas Belt, author of *The Naturalist in Nicaragua,* wrote:

We find in the Teo Amoxtli, as translated by the Abbé Brasseur de Bourbourg, an account of the overwhelming of a country by the sea, when thunder and flames came out of it, and "the mountains were sinking and rising." Everywhere throughout America there are traditions of a great catastrophe, in which a whole country was submerged, and only a few people escaped to the mountains; and the Spanish conquerors relate with wonder the accounts they found amongst the Indians of a universal deluge. Amongst the modern Indians the traveler, Catlin, relates that in one hundred and twenty different tribes that he had visited in North, and South, and Central America, "every tribe related, more or less distinctly, their tradition of the deluge, in which one, or three, or eight persons were saved above the waters on the top of a high mountain."[1]

Evidently people, except perhaps those living the most unruffled lives in the most sedentary societies during the most peaceful eras, have been aware for a long time that the world around them can change, often drastically and suddenly. Even religious traditions, which depend for their transmission on unchanging beliefs and values, admit the fact of change. In the Jewish morning service, for example, there are the words "in Your goodness You renew the work of creation every day, constantly." On a less pleasant note are the words in Deuteronomy, chapter 11: "Take heed to yourselves, lest your heart be deceived, and you turn aside and serve other gods, and worship them; and the anger of the Lord be kindled against you, and He shut up the heaven so that there shall be no rain, and the ground shall not yield her fruit; and you perish quickly from off the good land which the Lord gave you." In the New Testament there is also plenty of discussion of change, especially the now familiar verse from the book of Revelation, chapter 16: "And he gathered them together into a place called in the Hebrew tongue Armageddon," followed a few chapters later by the statement, "And I saw a new heaven and new earth: for the first heaven and the first earth were passed away; and there was no more sea."

None of these changes that I have listed and that various people have believed in for a long time—a deluge, the renewal of creation, the shutting of the heavens, the end of the sea, and the passing away of our familiar heaven and Earth themselves—are supposed to be the work of human beings. As I indicated, the idea that people, without divine intervention, can work vast changes in the Earth, is comparatively new. I do not mean that nobody ever thought of it or thought it possible until now—of course they did—but not until the nineteenth and twentieth centuries did it become a dominant idea of any major society.

Naturally there is no single moment when such a great transition in human thinking takes place. Nevertheless, if I had to pick one day most symbolic of this change I would choose November 17, 1869, the day on which Napoleon's dream of seventy years earlier came true and the Suez Canal was opened to traffic, shortening at one stroke the route from England to India by 6000 miles and forever changing the functional geography of the Old World. All one has to do to relive the feeling of new-found power associated with this event is to listen to the "Grand March" from *Aida,* the opera that was commissioned for the Suez Canal dedication and that was first performed at the Cairo Opera House in 1871. This is the musical personification of the dream of Suez, the dream of power and control, of progress, which has increasingly dominated our thoughts since that time.

The new idea emerged slowly—slowly enough so that, by the time it became commonplace, many of its side effects and problems were already anticipated. In a way, the problems of the dream of Suez were partly foreseen in Christopher Marlowe's *Doctor Faustus,* which appeared at the close of the sixteenth century. But the problems were more directly anticipated just four years before the opening of the Suez Canal, in 1865, at the end of the American Civil War, when a book by George P. Marsh was published in New York. Marsh, a diplomat, student of linguistics, and geographer, entitled his work *Man and Nature; or Physical Geography as Modified by Human Action.*[2] On the title page is a quotation from a famous sermon by the Protestant minister Horace Bushnell:

Not all the winds and storms, and earthquakes, and seas, and seasons of the world, have done so much to revolutionize the earth as MAN, the power of an endless life, has done since the day he came forth upon it, and received dominion over it.

There then follow 549 pages describing major changes that people have worked on the Earth and its creatures and the consequences—good and bad—of these changes. In this first great work of modern conservation, Marsh made it plain that there is no longer any place on Earth free of human influence. He wrote about large-scale human operations that "interfere with the spontaneous arrangements of the organic or inorganic world," and he suggested "the possibility and the importance of the restoration of disturbed harmonies and the material improvement of waste and exhausted regions." Man is, Marsh observed, "both in kind and degree, a power of a higher order than any of the other forms of animated life, which like him, are nourished at the table of bounteous nature."

Marsh believed an idea that had first been suggested by Charles Babbage nearly thirty years earlier, that *anything* a human being or animal does, whether it is just to make a sound or to take a step, leaves a permanent, never-dying impression on the processes and physical substrate of the Earth.[3] Because humans exert so much more power than animals, the lasting impressions we make on the planet are much greater than theirs. Consequently, whenever we make some permanent change in the Earth, disturb some harmony, it is our responsibility to make some other change that will restore the harmony.

Looking at Marsh's work, we can see that, not only was he the founder of the idea of conservation, at least in this country, but also he was the founder of a particular school of conservation, the interventionist, managerial school, which takes our disruptive presence in the natural world for granted and makes no judgments about that presence.

By 1920, it was possible for the great Scottish naturalist James Ritchie to write in his monumental book, called *The Influence of Man on Animal Life in Scotland:*

The animal life or fauna of a country is no fixed unit of occupation, established and unchanging, but, endowed with the plasticity of life, it carries in itself the imprints of many influences which have played upon it throughout the ages. . . .

So sweeping are the changes wrought by Man and so swift are they in their action that they obscure and almost submerge the slow march of the other processes of nature, and this difference in degree, associated with Man's purposefulness, almost inevitably leads to a sharp distinction being drawn between nature and man . . . but Man himself is still "Nature's insurgent son."

A little further on, Ritchie continues the theme:

Man has been described from one point of view as an instrument of destruction and from another as a creative agent. The truth of the matter as regards his relations with Nature is that he is neither all in all a destroyer nor a creator, but exercises his powers mainly as a transformer and a supplanter. Wherever he places his foot, wild vegetation withers and dies out, and he replaces it by new growths to his own liking, sometimes transformed by his genius for his own use. Where he pitches his tents and builds his cities, wild animals disappear, and woodlands and valleys where they sported are wrested from their prior owners and given over to the art of agriculture and to animals of man's own choosing, as well as to a host of camp followers, which attach themselves to his domestication whether he will or no. Intentionally and unintentionally, directly and indirectly, man transforms and sup-plants both animal and vegetable life. Some animals he deliberately destroys, some he deliberately introduces, and the characters of some he deliberately transforms by careful selection and judicious interbreeding. Other animals find his presence un-congenial and gradually dwindle in numbers or disappear, while others are encour-aged by his activities to increase in numbers, sometimes even to his own confounding.[4]

Notice how dispassionate this is. Human presence in the world is a fact; nature is bound to change and adjust in response to our presence, sometimes in ways that we plan, often in ways that we cannot predict or control. This was early in the twentieth century; more recently, as I discuss later, it has become easier to forget the part about not predicting and not controlling, and this technological optimism has even influenced many conservationists.

But meanwhile, as the Marshes, the Gifford Pinchots, the Teddy Roosevelts, and the James Ritchies were taking a human presence in nature for granted, another school of conservation thought was forming, setting the stage for a confrontation that continues today. In April of 1868, when Marsh's book had been out three years and while de Lesseps was finishing the Suez Canal, a Scottish-born inventor, wanderer, and naturalist named John Muir was beginning his first visit to California. Here is what he wrote about it ten years later:

Arriving by the Panama steamer, I stopped one day in San Francisco and then inquired for the nearest way out of town. "But where do you want to go?" asked the man to whom I had applied for this important information. "To any place that is wild," I said. This reply startled him. He seemed to fear that I might be crazy, and therefore the sooner I was out of town the better, so he directed me to the Oakland ferry.

Guiding himself by means of a pocket map, Muir headed east, toward the Yosemite Valley. He continued his account:

Looking eastward from the summit of the Pacheco Pass one shining morning, a landscape was displayed that after all my wanderings still appears as the most beautiful I have ever beheld. At my feet lay the Great Central Valley of California, level and flowery, like a lake of pure sunshine, forty or fifty miles wide, five hundred miles long, one rich furred garden of yellow *Compositae*. And from the eastern boundary of this vast golden flower-bed rose the mighty Sierra, miles in height, and so gloriously colored and so radiant, it seemed not clothed with light, but wholly composed of it, like the wall of some celestial city.[5]

Muir could not wait to leave San Francisco; the only kind of city in which he could thrive was a city without people, the "celestial city" of the High Sierra. It was Muir, more than anyone else, who established the other great school of conservation, in which conservation became synonymous with the protection and preservation of wilderness, the erection of fences to protect nature from all but the most passive of humankind.

In the years that have intervened since the days of Marsh and Muir, the two schools of conservation, the school of wise management and the school of wilderness preservation, have persisted, with surprisingly few changes that are usually brought about by the passage of time. But there have been some changes. The management conservationists have been bolstered by the ranks of thousands of hunters, trappers, and fishermen. The wilderness conservationists get the frequent assistance of the rapidly growing animals' rights movement. And neither school has retained its original purity. Conservation management is frequently described as "multiple use"—a helpful concept except when multiple use means that a natural habitat is divided up among the foresters, the mining engineers, the petroleum geologists, the hunters, the ski resort operators, the dam builders, the restauranteurs, and the drivers of off-road vehicles, with the native species of plants and animals getting any part of the habitat that may happen to be left over.

The wilderness protectors have lost their virginity, too. No longer does it suffice to protect the Earth for the reasons John Muir gave: that the wilderness is a "Godful" place and beautiful, a "celestial city." Now we learn that the reason to save the wilderness is because it is a potential source of new drugs to cure cancer, of hydrocarbons and fuel oils from plants, of natural rubber, of genes for insect resistance of crop plants, of plants whose leaves indicate the presence of gold in the soil, or of manatees—sea cows—to eat stream-clogging water hyacinths.[6] Not that these are not valid possibilities; yet it is hard in this age of near-universal selfishness, materialism, and unease to read book after book of practical reasons for saving wilderness without feeling a twinge of regret for the passing of the priceless, uncorrupted wilderness of John Muir. Now, as with everything else, wilderness has its price. It is also difficult to imagine wilderness remaining wilderness after the drug companies, hydrocarbon refiners, gold miners, manatee catchers, and rubber planters have finished with it. In an ideological war it is always dangerous to adopt the rationale of the enemy. But perhaps I quibble.

In all this, one fact remains absolutely clear. Marsh was right: human beings can have and increasingly are having a profound and lasting effect on the natural world. Few people comprehend the magnitude of that effect. Large numbers alone without a recognizable context have little meaning. According to the Worldwatch Institute, thanks to agricultural practices the Earth is now losing 23 billion tons of topsoil annually, over and above new soil formation.[7] But what does 23 billion tons mean? Considering the size of the Earth, is it a lot or a little?

I prefer a different statistic: on April 3, 1924, at the Agricultural Research and Education Center in Belle Glade, Florida, a 9-foot, graduated concrete post was sunk straight down through the rich muck soil of this drained part of the Everglades until it hit bedrock and the top of the post was at ground level. The area was fenced off. By April, 1979, fifty-five years later, the forces of oxidation working on the soil that had been drained for an inappropriate kind of agriculture had lowered the ground level so that 5 feet of the pole was exposed.[8] In this protected spot less than half of the soil that now produces (in nearby parts of south Florida) so much of our winter vegetable harvest remains. In some places in Florida, that soil is now only 12 inches over the limestone bedrock, and it is becoming impossible to farm. One of the largest growers in Florida has recently purchased a vast tract of agricultural land in Australia. Evidently these are statistics with a practical meaning.

In the Soviet Union, the Aral Sea in central Asia, a huge body of water with an average depth of 50 feet in 1965, is shrinking because the rivers that once fed it with fresh water have been entirely diverted for agricultural irrigation. By the year 2000, it is projected that the Aral Sea will no longer exist—it will be replaced by a salt marsh.[9]

In El Salvador in 1961, there were approximately a half-million acres of forest and woodland. By 1978, the total forested acreage of El Salvador was close to zero. It had all been turned into pasture for beef cattle for fast food hamburgers in the United States.[10]

Naturally the worldwide extinction of species parallels the scale of these sorts of incredible habitat destructions and alterations. In 1970, exasperated by some people who were claiming that extinction was a normal process and therefore nothing to worry about, I compared the modern extinction rate among most groups of mammals with the extinction rate that occurred during the last great die-off, in the ice ages of the late Pleistocene Epoch. The extinction rate of 1970 was, conservatively calculated, about a thousand times greater. And it is worse now, only fourteen years later. Recent calculations by various ecologists are more sophisticated than mine but not more encouraging. Florida State University's Daniel Simberloff has estimated that by the time the world's primary tropical forests are two-thirds cut—perhaps two to three decades from now—as many as 625,000 species will have become extinct. Other estimates are considerably higher. Only 1.7 million species of plants and animals have been named to date.[11]

Where are we headed? We near the end of the twentieth century confronted by the loss of many of the species in whose company we began the century. In major group after group of plants and animals we find a significant percentage, sometimes a majority, occasionally all of the species endangered. Among the less-studied groups, we have undoubtedly lost species that we never even got a chance to name or even to recognize. In the face of this danger, what kinds of remedies can we apply? Only two, in fact—the ones we have been developing since the days of Marsh: protection or fencing off and conservation management. Neither, I am afraid, is up to the task.

Fences and gamekeepers to keep people out of protected areas may have worked in the days of King Henry the Eighth; they do not work now. I think of the Hutcheson Forest of Rutgers University, a 60-acre piece of pristine, precolonial forest in central New Jersey. Never cut, never plowed. The last major fire in 1711. Used only for unobtrusive ecological research. The public kept out except for closely guided tours on a single narrow trail, once every two to four weeks. And what is it like? There are great gaps in the canopy, crisscrossed by trailing vines of grape, poison ivy, and bittersweet, a sure sign of severe disturbance. Exotic, nonnative species are everywhere: Norway maple, *Ailanthus* (the tree of heaven), Japanese honeysuckle, the egg cases of gypsy moths. You can fence out people, but you cannot fence out their effects. And this is a place where there are no rhinoceroses and elephants to attract the horn and ivory poachers, no exotic, tropical orchids to tempt unscrupulous plant collectors; it should be easy to protect. But alien introduced pests, acid rain, ozone, insecticide residues, drifting herbicide, heavy metals, atmospheric particulates—these effects and creations of our society can be anywhere and everywhere on Earth.

Not only the effects of people but also the people themselves are more pervasive and invasive than previously. There are more people than there ever were before, many of them with leisure and money, and they have scubas and snowmobiles. Where can people be kept from overrunning nature? Remote islands and top-security military bases are fairly well protected, yet that is only a small percentage of the Earth's surface.

There is another problem with protection. In ecosystem after ecosystem, ecological studies are beginning to show that a protected oasis in a sea of development loses species quickly. How big does a park or preserve have to be to save its species?[12] This question is only now being asked in different parts of the world, but the answer is already becoming familiar: as big as you can get, and even then the results will never be entirely satisfactory.

If protection is a weak reed, I am afraid that active management and intervention are not much better, despite our technological self-confidence. Consider as an example of this type of conservation the preservation of animal species in zoos or special breeding facilities and the preservation of agricultural plant varieties in seed banks. William Conway, the director of the New York Zoological Society, and George Rabb, director of the Chicago Zoological Park, both effective conservationists, computed in 1980 that the care of the 750 Siberian tigers now in the world's zoos will cost $49 million just to maintain them for twenty years until the year 2000. Five hundred gorillas will cost $47 million. If we saved just 2000 selected species out of the tens and probably hundreds of thousands that are endangered, it would cost $25 billion by the year 2000, said Conway, who, I should point out, was advocating the effort.[13] Moreover, this is only the economics of it—never mind the genetic worries about whether the species will become irreversibly changed in captivity or the ecological problem of the disappearance of their native habitat while they are in captivity.[14]

Seed banks pose similar problems. Expensive, ecologically and evolutionarily nonsensical, and vulnerable to the loss of entire collections through accidents such as power failures, through sabotage, or through bureaucratic misunderstandings, seed banks are inadequate to preserve more than a small fraction of even the recorded varieties of crop plants. At Kew Gardens, in England, it is estimated that the new acquisitions are just balancing the seeds that are being lost from storage. The situation at the U.S.'s major seed bank, at Fort Collins, Colorado, appears to be even worse.[15] This is not an encouraging record.

Beyond zoos and seed banks, which after all, do save species in spite of the difficulties, there is the ultimate dream of management-oriented conservation: genetic engineering. We will make new species to replace the old ones, some say.

Nachmanides, Rabbi Moses ben Nachman, a thirteenth-century Spanish biblical commentator, observed that because Noah's ark, as described in the Bible, was not even remotely big enough to hold representatives of every animal species, God must have worked a miracle to get them all inside and keep them alive. That kind of realistic and practical thinking is not popular today, with our technological euphoria in full swing. "You want your species back? We'll recreate them," the technophiles promise, ignoring all the accumulated knowledge of biology, physics, and information theory simultaneously. Recreating from scratch even one species would be a feat that would make the stuffing of animals into Noah's ark simple by comparison. Barring a miracle, genetic engineering as a conservation tool is simply not worthy of consideration as a serious option.

Before going on, I want to stress that we must continue and redouble all our legitimate conservation efforts of both the protection and the management varieties—the parks, preserves, and wilderness areas, the seed banks, the zoos, the endangered species facilities, and the aquariums. They will not be adequate to save the world's endangered species if things continue on their present course, but suppose, as is likely, that our course changes. What then? True, the state of affairs might get worse, but it also might get better, and in that event any interim conservation that we have achieved will be of value.

Let us examine some of the possible types of change by projecting trends and processes that are already clearly present. This is not as adventurous an approach to prediction as science fiction or even conventional futurology; however, I hope that it is more solid and realistic. At the outset, I want to say that I believe that *the ultimate success of all conservation will depend on a revision of the way we use the world in our everyday living when we are not thinking about conservation.* If we have to conserve the Earth in spite of ourselves, we will not be able to do it.

Let me give an example—a modern parable of conservation. In the Papago Indian country of Arizona's and Mexico's Sonoran Desert, described so beautifully in Gary Nabhan's book *The Desert Smells Like Rain,* there are two similar oases only thirty miles apart.[16] The northern one, A'al Waipia, is in the U.S. Organ Pipe Cactus National Monument, fully protected as a bird sanctuary, with no human activity except bird-watching allowed. All Papago farming, which had existed there continuously since prehistory, was stopped in 1957. The other oasis, Ki:towak, over the border in Mexico, is still being farmed in traditional Papago style by a group of Indian villagers.

Visiting the oases "on back-to-back days three times during one year," Nabhan, accompanied by ornithologists, found fewer than thirty-two species of birds at the Park Service's bird sanctuary but more than sixty-five species at the farmed oasis. Asked about this, the village elder at Ki:towak replied:

I've been thinking over what you say about not so many birds living over there anymore. That's because those birds, they come where the people are. When the people live and work in a place, and plant their seeds and water their trees, the birds go live with them. They like those places, there's plenty to eat and that's when we are friends to them.

And that is when conservation becomes reality, when people who are not actively trying to be conservationists play and work in a way that is compatible with the existence of the other native species of the region. When that happens—and it happens more than you may think—the presence of people may enhance the species richness of the area, rather than exert the negative effect that is more familiar to us. With this in mind, I can consider some of the alternatives we may face in the days ahead.

First and most obvious, is the possibility—not a likely one, I think— that during the next twenty-five to fifty years things will go on pretty much as they have in the recent past since the 1950s. In other words, there will be more people, there will be more industrialization, there will be more urban growth, there will be more standardization, there will be more corporate conglomeration and bigger organizations, there will be more power-oriented consumer goods, there will be more tourism, there will be bigger weapons budgets for more elaborate weapons, there will be more advertising and image making, there will be less room for personal eccentricities, and there will be more mechanization of agriculture and chemical farming. There will also be no major cataclysmic upheavals, either economic or military, to disturb this pattern of life.

If this is what happens to us during the next half-century, then it becomes a fairly easy job to predict the fate of the animal and plant species on Earth. In 1970, in my textbook *Biological Conservation,* I had a section that I titled "Characteristics of Endangered Species."[17] In this, there was a table with two columns, one labeled "Endangered" and the other labeled

"Safe." In the table I listed the sorts of characteristics that might put a species at high or low risk of extinction, and I gave contrasting examples using related species of animals. For example, endangered species of animals are likely to be of large size, such as the cougar; safe species are likely to be small, such as the wildcat. Species that have a restricted distribution—an island, a few bogs, a desert watercourse—are often endangered; the Puerto Rico parrot is an example. Species with wide distribution, such as the yellow-headed Amazon parrots, are at least comparatively safe. Species that are intolerant of the presence of humans, the grizzly bear for example, are endangered; tolerant species, such as the black bear, are safe. Species with behavioral idiosyncrasies that are not adaptive in urban and suburban areas do poorly: The redheaded woodpecker flies from one tree to the next in a low, swooping arc that often brings it into the path of oncoming cars. On the other hand, some species have, by chance, behavioral patterns that suit them for coexistence with us. The burrowing owl is tolerant of noise and has a kind of flight that lets it evade oncoming objects. The last I heard, they were doing quite well living between the runways of the Miami International Airport. This is a sample of some of the characteristics I listed in the table. (In 1974, John Terborgh, at Princeton, published an interesting paper on what he called "extinction prone species," with some similar descriptions.[18])

Having listed the characteristics of endangered species of animals, I went on to construct a hypothetical "most endangered animal." It is

a large predator with a narrow habitat tolerance, long gestation period, and few young per litter. It is hunted for a natural product and/or for sport, but is not subject to efficient game management. It has a restricted distribution but travels across international boundaries. It is intolerant of man, reproduces in aggregates, and has nonadaptive behavioral idiosyncrasies. Although there probably is no such animal, this model, with one or two exceptions, comes very close to being a description of a polar bear.

Conversely, if you take the opposite characteristics—small size, herbivorous diet, high fecundity, wide distribution, and so forth—you get

a composite picture of the "typical" wild animal of the twenty-first century—some of the most familiar existing approximations would be the house sparrow, the gray squirrel, the Virginia opossum, and the Norway rat.

We could also include in that list the common pigeon or rock dove, the familiar species of domestic cockroach, the feral house cat, and such creatures as the rapidly spreading Eastern canid, which is part dog and part coyote, with maybe a little bit of wolf mixed in. Among the plants, we will see more of *Ailanthus,* which thrives in areas covered with asphalt and concrete, ragweed, which does best in recently disturbed soils, and the tall reed grass *Phragmites,* which loves wet spots, regardless of how much air and water pollution there may be in the neighborhood.

These are the jolly companions we can expect if the world continues much longer on its present course. The tropical forests will be gone, along with their myriad animal species; the temperate forests will be going. There will continue to be rhinoceroses in zoos, showy orchids in botanical gardens, and slowly or rapidly deteriorating, poorly representative collections in seed banks of what is left of the human agricultural heritage: a few varieties of African upland rice, more varieties of wheat, a smattering of beans. The fruit trees will fare even worse, although I suppose we will have the awful Red Delicious apple around forever, as a perpetual reminder of our sins.

That is the future as I see it, if we simply extend what is happening now. No conservation efforts, either of the protective or managerial sort, will reverse this basic trend toward extinction, although there will be a few victories here and there, now and then. I do not mean to imply that the species that will be left will be only the weeds, the pests, and the vermin, although these are certainly the species that do best in damaged ecosystems. There will still be salutary species around but far fewer of them than there are now. We are selecting for the tough species, the resilient species, the species of upheaval, and these tend to be the ones that we do not like and that are not good for us. As it says in a superficially simpleminded statement in Deuteronomy, chapter 30, "choose life, that you may live." Conversely, if we continue to choose a society that survives through destruction and death, we should not be surprised if destruction and death are what we get.

Before leaving this particular view of the future and going on to another alternative, I must deal with one part of modern life that many people feel will completely change the course of future events and invalidate all projections based on present tendencies. I refer to the so-called information revolution. This revolution, made possible by computers and their associated software, has developed in three areas: (1) the storing, retrieving, and manipulating of numerical information; (2) communications; and (3) the redesigning of organizational structures, which is just beginning. To what extent do these developments modify the prognosis for species in the world of the future? Will our mastery of information and information handling enable us to avert the biological problems caused by the rest of our technology?

The late Father Teilhard de Chardin foresaw the day when human consciousness would flow together into one great unified layer, becoming part of the "noösphere," enveloping the Earth in its collective intellectual and spiritual wisdom.[19] Have we moved, are we moving in that direction? I think not.

Those who predict major, even unlimited benefits from the information revolution ignore the existence of certain fundamental limits inherent in it: the limit to the value and usefulness of pure information is one; the limit to the quality of information available is another; the limit to the malleability and perhaps the complexity of human social structures is the last. These limits interact in infinite ways to limit severely the possibilities of the information revolution, and we are already beginning to see that. I can give only a few scattered examples.

What I am talking about goes beyond the GIGO (garbage in, garbage out) principle. We are foundering in a sea of information. In one of the most profound and funniest science fiction stories of Stanislaw Lem, the heroes, Trurl and Klapaucius, are trapped inside their spaceship in a remote junkyard corner of space by a pirate named Pugg, a hideous robotic monster with a Ph.D. Pugg wants information even more than gold, so Trurl and Klapaucius design for him a variant of the perpetual motion machine, a gadget that generates random facts about the universe at incredible speed and then selects and prints out those that are true.

The tiny diamond-tipped pen shivered and twitched like one insane, and it seemed to Pugg that any minute now he would learn the most fabulous, unheard-of things, things that would open up to him the Ultimate Mystery of Being, so he greedily read everything that flew out from under the diamond nib . . . the sizes of bedroom slippers available on the continent of Cob, with pompons and without . . . and the average width of the fontanel in indiginous stepinfants . . . and the inaugural catcalls of the Duke of Zilch, and six ways to cook cream of wheat . . . and the names of all the citizens of Foofaraw Junction beginning with the letter M, and the results of a poll of opinions on the taste of beer mixed with mushroom syrup. . . .

And it grew dark before his hundred eyes, and he cried out in a mighty voice that he'd had enough, but Information had so swathed and swaddled him in its three hundred thousand tangled paper miles, that he couldn't move and had to read on about how Kipling would have written the beginning to his second *Jungle Book* if he had had indigestion just then, and what thoughts come to unmarried whales getting on in years and all about the courtship of the carrion fly . . . and why we don't capitalize paris in plaster of paris.[20]

And many more things that I won't include. Meanwhile, Trurl and Klapaucius escape, leaving Pugg to his hideous punishment, which is also recorded somewhere in the many miles of tape coming out of the information machine.

Another problem with the information revolution is what, for lack of a better phrase, I call the real goods problem. We cannot eat information, we cannot wear it, and information will not take out the garbage at night. As an increasing percentage of society becomes involved with the manipulation of information, much of it worthless, fewer people are left to concern themselves with the real goods and services needed for carrying on life. It can be argued that the information manipulators are increasing the efficiency of everyone else. But again there is a limit to what can be accomplished by increasing the efficiency of designs and processes, and efficiency, itself, is not always desirable.

What I see happening in the information revolution is a massive and complicated trade-off, a trade-off that is sometimes beneficial and sometimes harmful, but which is certainly not leading us toward the paradise promised by the revolution's marketers. A good illustration of this is provided by the telephone. Telecommunications technology has explored in recent years with satellite relays, microwave relays, fiber optics, computer-controlled traffic flow, electronic voice synthesizers, computer monitoring of operator efficiency and phone charges, computerized information storage, wireless telephonic systems, and much more. And how has telephoning changed for us? What is it like to make a phone call, to dial "information," compared with what it was like twenty years ago? What about the speed and cost of repairs if your phone is broken? What about the monthly bill? This is what business people call the bottom line. Twenty years ago, a call to the information operator was free. Now, in the midst of the information revolution, you can listen to the electronic voice telling you your number—better get it by the second time—and you can think of how big your phone bill is going to be at the end of the month and wonder how long it will take for the American phone system to deteriorate to the level, say, of Egypt's. And when you finally get through to your old friend in Decatur, you can talk about the weather out there and who has died and who has been divorced, just as you might have done if you had been phoning in 1964, or 1934 for that matter.

So I do not think that the information revolution is likely to modify the sad forecast for the fate of species if things continue on as they have in the recent past. As we speed along the express track of modern technological inventiveness, headed toward a clear and brilliant future, we hear a roar and feel the pull as another train rushes by in the opposite direction. Look quickly, and we will see ourselves seated there, too, on the express called the Reality of Life, headed not into the sunrise but into the storm and the gathering darkness.

But suppose that things do not continue on those tracks. Suppose that things change. There are three possible changes I want to consider, none of them at length.

The one change that nobody likes to mention and everybody things about is the change that would be brought on by nuclear war. I have no interest in speculating about the kind of creatures, and there would be some, that would survive a nuclear war. The only thing that concerns me is that there would be no human society left that would be fit to take cognizance of them. I am not endorsing the nuclear winter hypothesis of Sagan, Ehrlich, and a number of other distinguished and courageous scientists.[21] They are probably correct, but there may be some scientific variable they have overlooked, which if included in the analysis would change the results. My point is that after even a so-called limited nuclear war, if it is not nuclear winter that delivers the coup de grace to this planet's ecosystems, then it will be something else. If your automobile is not running too well, you can, if you wish, shoot an armor-piercing grenade into it. It is conceivable that the explosion will rearrange things in such a way that the engine will get a good tune-up and valve job. It is conceivable. But if you count on that happening, you deserve what you will get.

Another change that people do not like to speak of—not as bad as nuclear war, but certainly no fun—is the change that would be brought on by global economic collapse (assuming no nuclear consequences), a collapse that would probably decrease international trade, trigger the disintegration of many multinational corporations and other overstuffed, subsidized superorganizations, end the modern welfare state, bring about massive famines, greatly increase unemployment in the industrialized nations, all but eliminate luxury goods and exceedingly complex manufactures, including many complex military weapons, increase local wars, and cause a great proliferation of regional economic systems.

I believe that such an economic collapse would have two different and opposite effects on world faunas and floras. In countries such as India, Bangladesh, Mexico, and some of the more densely populated of the African nations, where natural resources, especially trees, are already seriously depleted, little of the native flora and fauna would survive. As export industries and cash crops failed, unprepared and unsupplied local agricultural communities would have to feed and provide cooking fuel for large urban populations, an impossible burden on the land. Anyone who flies over Mexico today will see what an obvious prediction this is. On the other hand, widespread economic collapse would also bring about the cessation of many disruptive processes. No longer would Japanese industry and companies like Weyerhauser and Georgia-Pacific be cutting the last remaining primary forests of Australia and the South Pacific at today's staggering rate. No longer would Volkswagen and Liquigas and the like be using defoliants or napalm or bulldozers to turn Amazon rain forest into short-lived cattle ranches for the fast food trade. No longer would there be a strong market for pet Hyacinthine Macaws and Palm Cockatoos, at $5000 or more apiece. So, although a global economic breakdown would strain the ecosystems and species of some places past the breaking point, it would give ecosystems and species in other places a new lease on life. And in a few areas these contradictory processes would be happening side by side.

The last possibility of change I want to mention is both noncatastrophic and, to a conservationist, very hopeful. This possibility is the transformation of the dream of progress from one of quantity, production, consumption, monumental waste, and the idiot's goal of perpetual growth to one of quality, equilibrium, durability, and stability—in part based on the inventive imitation of natural systems. The latter is also a dream of progress because it, too, like the chemical electronic life we are now striving for, would have to be invented and built by bright people. It does not yet exist.

Of course we already have had hints of what this kind of change would be like: the city of Florence in the Renaissance; the *chinampas,* or swamp gardens, that were the glory of fifteenth-century Aztec farming; the hedgerows of post-Elizabethan England; old Jerusalem and the terraces of the nearby Judean Hills; and the ingenious and intricate multicrop gardens of tropical western Africa, to name a few.

Such a change in world view would involve much more than agriculture, but agriculture and its side effects are now the major cause of the worldwide loss of species, so this is all I will mention. Imagine an agriculture with field edges colonized by skillfully placed, useful weeds including the wild relatives of the crops growing in the fields. Imagine Midwestern grain farms modeled after the most soil-building, stable, drought-resistant, insect-resistant, and productive grassland ecosystem known—the prairie itself. This work has been started by the Jacksons, at the Land Institute, in Salina, Kansas; it will thrive—or we won't.[22] Imagine combined wind- and diesel-powered fish-hauling vessels, designed to minimize consumption of precious fuel, in third world maritime nations. This has been started by the Todds, at Woods Hole, Massachusetts.[23] It too will thrive—or we won't. Imagine farms with their owners living on them, farms small enough so that the farmer knows every square foot of his or her property. The Amish do it brilliantly.[24] Others will have to learn.

Figure 7.1

William Blake, from the *Songs of Innocence and Experience*. A shepherd guards his flock with care in the shade of a spreading tree. It is dawn of a summer day; we are looking east across the hills. The sheep appear well fed and content. A bird of paradise flies up overhead, and other birds are visible. When part of nature is used in a reverent fashion, the rest of nature thrives. Both wildness and wise use have their place in a conserved world. Photograph courtesy of Houghton Library, Harvard University.

How might such a change come about? It might arise from the chaos following economic collapse; most of us hope that is not the way. It might result imperceptibly from many small, forced adjustments caused by scarcity and breakdown within the present system. This is what was predicted by Warren Johnson in his book *Muddling Toward Frugality*.[25] And it might happen because of one of those great and comparatively sudden transitions in human attitude toward the world and society that have happened occasionally in the past and that religion has more to say about than science.

To conclude, we know what our distant ancestors did not know: we can change the world utterly and completely. In the face of this, neither of the classical varieties of conservation, protection or management, can by itself save the world's fauna and flora. Nevertheless, they are both vitally necessary as a holding action to save what can be saved until such a time, soon or far off, when humanity adopts a way of life in harmony with other life forms on the planet.[26] If and when that happens, we will necessarily be living in a world characterized by ecological management—stewardship is an appropriate description—so I judge that the managerial type of conservation will predominate. Yet even in such a world there will always remain unknowns and imponderables, many things beyond our control and understanding, and it is for this reason that simple protection, which makes no assumption of human omniscience, will continue to be an occasional choice of a society concerned with the fate of other species.

Yet for all this, it cannot be denied that, before we can begin the days of stewardship, we must first end and leave behind the age of exploitation, which is still very much with us. We can only pray that the passage comes soon and that it is a peaceful one—soon, while there is still much to safeguard; peaceful, so that we have a fair chance to prove our worth as stewards.

Notes

1. Thomas Belt, *The Naturalist in Nicaragua* (London: John Murray, 1874), 271–272.

2. George P. Marsh, *Man and Nature; or Physical Geography as Modified by Human Action* (New York: Charles Scribner, 1865).

3. Charles Babbage, *The Ninth Bridgewater Treatise: A Fragment,* new impression of the second edition (London), 1838 (London: Frank Cass, 1967), 108–119.

4. James Ritchie, *The Influence of Man on Animal Life in Scotland: A Study in Faunal Evolution* (London: Cambridge University Press, 1920), v, 5.

5. John Muir, *The Yosemite* (Garden City, New York: Doubleday Anchor, 1962), 1–2. Originally published in 1912.

6. Margery L. Oldfield, *The Value of Conserving Genetic Resources* (Washington, D.C.: National Park Service, 1984).

7. Lester R. Brown et al., *State of the World: 1984* (New York: Norton, 1984), 61–62.

8. Robert L. Tate, III, "Microbial oxidation of organic matter of histosols," in *Advances in Microbial Ecology,* M. Alexander, ed. (New York: Plenum, 1980), vol. 4, 169–201.

9. Marshall I. Goldman, *The Spoils of Progress: Environmental Pollution in the Soviet Union* (Cambridge, Massachusetts: The MIT Press, 1972), 217.

10. Lester R. Brown et al., *State of the World: 1984,* 78. Val Plumwood and Richard Routley, "World rainforest destruction—the social factors," *The Ecologist* 12(1) (1982), 4–22.

11. Roger Lewin, "No dinosaurs this time," *Science* 221 (1983), 1168–1169. Daniel Simberloff, "Mass extinction and the destruction of moist tropical forests," *Journal of General Biology (USSR),* in press. Edward O. Wilson, "The biological diversity crisis," *Bioscience* 11 (1985), 700–706.

12. Roger Lewin, "Parks: How big is big enough?" *Science* 225 (1984), 611–612.

13. William Conway, "Zoos: Future directions," *Animal Kingdom* (April-May 1980), 28–32.

14. Paul Ehrlich and Anne Ehrlich, *Extinction: The Causes and Consequences of the Disappearance of Species* (New York: Random House, 1981), 207–233. Robert M. May, "Inbreeding among zoo animals," *Nature* 283 (1980), 430–431.

15. *Conservation Foundation Letter,* "Medicinal plants need extensive safeguarding" (November 1982). O. H. Frankel, "Genetic conservation in perspective," in *Genetic Resources in Plants—Their Exploration and Conservation,* O. H. Frankel and E. Bennett, eds. (Philadelphia, Pennsylvania: Davis, 1970), 469–489. Peter H. Raven, "Ethics and attitudes," in *Conservation of Threatened Plants,* J. B. Simmons et al., eds. (New York: Plenum, 1976), 155–179. Marjorie Sun, "Fiscal neglect breeds problems for seed banks," *Science* 231 (1986), 329–330.

16. Gary Nabhan, *The Desert Smells Like Rain: A Naturalist in Papago Indian Country* (San Francisco, California: North Point Press, 1982), 89–97.

17. David Ehrenfeld, *Biological Conservation* (New York: Holt, Rinehart, and Winston, 1970), 127–131. See also David Ehrenfeld, *Conserving Life on Earth* (New York: Oxford University Press, 1972), 200–204.

18. John Terborgh, "Preservation of natural diversity: The problem of extinction prone species," *BioScience* 24 (1974), 715–722.

19. Pierre Teilhard de Chardin, "The antiquity and world expansion of human culture," in *Man's Role in Changing the Face of the Earth,* William L. Thomas, Jr., et al., eds. (Chicago, Illinois: University of Chicago Press, 1956), 103–112.

20. Stanislaw Lem, *The Cyberiad* (New York: Avon, 1976), 119–134.

21. R. P. Turco et al., "Nuclear winter: Global consequences of multiple nuclear explosions," *Science* 222 (1983), 1283–1292. Paul R. Ehrlich et al., "Long-term biological consequences of nuclear war," *Science* 222 (1983), 1293–1300.

22. Wes Jackson, *New Roots for Agriculture* (Lincoln, Nebraska: University of Nebraska Press, 1985).

23. John Todd, "The ocean pickup on the Spanish Main: The saga continues," *The Annals of Earth Stewardship* 2(2) (1984), 8–13.

24. Wendell Berry, "Seven Amish farms," in *The Gift of Good Land* (San Francisco, California: North Point Press, 1981), 249–263. W. A. Johnson et al. "Energy conservation in Amish agriculture," *Science* 198 (1977), 373–378.

25. Warren Johnson, *Muddling Toward Frugality* (San Francisco, California: Sierra Club Books, 1978).

26. David Ehrenfeld, *The Arrogance of Humanism* (New York: Oxford University Press, 1981).

Beyond the Last Extinction

Environmental Organizations for Biodiversity

In the introduction to the resource section in the 1986 edition of *The Last Extinction,* we began by expressing the hope that this book would alert the reader to the reality and the seriousness of ongoing mass extinctions. Seven years later there can be little doubt that this message has spread and that today there is a much greater awareness of the need to preserve our planet's biodiversity. As mentioned in the preface to this second edition, Vice President Al Gore's book *Earth in the Balance* and Edward O. Wilson's compelling *The Diversity of Life* reflect this broader awareness in both political and academic circles. Similarly, the number of national and international governmental and citizens' groups devoted to environmental awareness and stewardship has burgeoned too.

We draw the following list of organizations from primarily one source, the *Conservation Directory* published annually by the National Wildlife Federation, 1412 Sixteenth Street, Washington, DC 20036-2266. This sampling of organizations involved in issues of species and habitat survival or related issues comes from the 1992 edition of the *Conservation Directory* and includes many new names and addresses, some of which complement the new chapter about whales that ecologist Norman Myers has contributed to this second edition.

The editors of this revised edition of *The Last Extinction* hope you will use this resource to take practical steps to get involved. Whatever your temperament, politics, income, or available time, you will almost certainly be able to find a group of similarly minded people. Don't wait—act now while there is still time to make a difference.

African Wildlife Foundation
1717 Massachusetts Avenue, NW
Washington, DC 20036
Focuses on practical steps toward the protection of African wildlife, including education, technical assistance, and conservation projects; works closely with governmental ministries.

Alliance for Environmental Education, Inc.
51 Main Street, P.O. Box 368
The Plains, VA 22171

A consortium of thirty organizations sharing the educational goal of developing personal and societal commitments to improving the quality of life.

American Association of Zoological Parks and Aquariums
Oglebay Park
Wheeling, WV 26003-1698

Promotes the multifaceted development of zoos and aquariums, with a special focus on conservation of wildlife and the preservation and propagation of endangered and rare species.

American Cetacean Society
P.O. Box 2639
San Pedro, CA 90731-0943

Organization dedicated to conservation, education, and research to protect marine mammals and their environment.

American Littoral Society
Sandy Hook
Highlands, NJ 07732

Professionals and amateurs interested in the study and conservation of coastal habitats and their wildlife.

American Minor Breeds Conservancy
Box 477, Suite 6
101 Hillsboro
Pittsboro, NC 27312

Coordinates conservation activities focusing on rare and endangered breeds of livestock.

Animal Welfare Institute
Box 3650
Washington, DC 20007

Active in improving conditions for laboratory animals, protecting endangered species, and educational activities.

Aquatic Conservation Network
540 Roosevelt Avenue
Ottawa, Ontario, Canada K2A 1Z8

Organization to improve communications in aquatic conservation and to enhance participation by amateur aquarists in scientifically guided programs of conservation and captive breeding.

Atlantic Salmon Federation, International Headquarters
P.O. Box 429
Street Andrews, New Brunswick
Canada EOG 2XO
Dedicated to the preservation and management of the Atlantic salmon and its habitat.

Caribbean Conservation Corporation
Box 2866
Gainesville, FL 32602
Focuses on endangered sea turtles and the protection of their nesting and feeding habitats throughout the Caribbean.

Center for Coastal Studies
P.O. Box 1036
Provincetown, MA 02657
Organization engaged in research, conservation, and education in the marine and coastal environment.

Center for Environmental Information, Inc.
46 Prince Street
Rochester, NY 14607-1016
Provides references and referrals on environmental issues.

Center for Marine Conservation, Inc.
1725 De Sales Street, NW
Suite 500
Washington, DC 20036
Organization to protect marine wildlife and their habitats and to conserve the oceans and coastal resources.

Center for Plant Conservation, Inc.
Box 299
St. Louis, MO 63166
A national network of twenty botanical institutions with the goal of preserving collections of threatened and endangered U.S. plant species.

Chesapeake Bay Foundation
162 Prince George St.
Annapolis, MD 21401
Organization to help save the Chesapeake watershed from New York and New Jersey down through the Bay, through land management programs, transportation plans, and educational awareness.

Clean Ocean Action
P.O. Box 505
Sandy Hook, NJ 07732
Coalition to improve the degraded waters off the coasts of New York and New Jersey through education, research lobbying, and citizen action.

Coolidge Center for Environmental Leadership
1675 Massachusetts Avenue, Suite 4
Cambridge, MA 02138-1836
Organizes educational programs, primarily for foreign graduate students, on issues of environment and development.

Cousteau Society, Inc., Headquarters
930 West 21st Street
Norfolk, VA 23517
Dedicated to the preservation of the oceans and the protection and improvement of life.

Defenders of Wildlife
1244 19th Street NW
Washington, DC 20036
Focuses on the integrity of natural wildlife ecosystems with the goal of preserving, enhancing, and protecting the natural abundance and diversity of wildlife.

Desert Botanical Garden
1201 North Galvin Parkway
Phoenix, AZ 85008
A repository for living collections of endangered desert plants.

Desert Fishes Council
Box 337
Bishop, CA 93541
Focuses on the status, protection, and management of the endemic fauna and flora of the North American desert ecosystems.

Desert Tortoise Preserve Committee, Inc.
Box 453
Ridgecrest, CA 93556
Acquires land sanctuaries and sponsors educational programs aimed at saving the desert tortoise of the American Southwest.

Earthwatch
Box 403N, 680 Mt. Auburn Street
Watertown, MA 02272
A clearinghouse for scientific research; individuals are invited to join scientific expeditions and thereby help to defray their cost. Focuses particularly on endangered cultures, animals, and habitats.

E. F. Schumacher Society
Box 76, RD3
Great Barrington, MA 01230

Conducts public seminars, lectures, and workshops on intermediate technology, regional land protection organizations, and small-scale cooperative enterprises.

Environmental Defense Fund, Inc.,
257 Park Avenue South
New York, NY 10010

Pursues responsible reform of public policy in the fields of energy and resource conservation, toxic chemicals, water resources, air quality, land use, and wildlife.

Environmental Policy Institute (see Friends of the Earth)

Conducts research, lobbying, and public education and works toward the organization of coalitions to address a broad range of natural resource issues.

Florida Audubon Society
460 Highway 436, Suite 200
Casselberry, FL 32707

Regional society engaged in a broad range of ecosystem and special conservation activities.

Forest Trust
P.O. Box 519
Santa Fe, NM 87504-0519

Seeks to protect and improve forest ecosystems and resources by providing land management strategies to the public and private landholders.

Friends of the Earth (merged with Environmental Policy Institute and Oceanic Society)
218 D Street SE
Washington, DC 20003

Committed to the preservation, restoration, and rational use of the earth through policy and education.

Fund for the Animals, Inc.
200 West 57th Street
New York, NY 10019

Focuses on the preservation of wildlife and saving endangered species and promotes humane treatment of all animals.

Global Tomorrow Coalition, Inc.
1325 G Street NW, Suite 915
Washington, DC 20005-3104

National alliance of organizations and individuals dedicated to education about the long-term significance of interrelated global trends.

Greater Yellowstone Coalition
Box 1874, 13 South Wilson
Bozeman, MT 59771

Network organization of people interested in the preservation of the Greater Yellowstone ecosystem and a greater appreciation of ecosystems in general.

Greenpeace, USA, Inc.
1436 U Street, NW
Washington, DC 20009

Advocates nonviolent direct action to confront environmental abuse.

Haplochromis Ecology Study Team
F. W. Witte
Zoologisch Laboratium
Postbus 9516 2300
Leiden, The Netherlands

Focuses on the preservation of fish species endemic to the Great Lakes of Africa.

Institute for Alternative Agriculture
9200 Edmonston Road, Suite 117
Greenbelt, MD 20770

Fosters the development and introduction of low-cost, low-chemical, and low-energy farming methods into American agriculture.

International Oceanographic Foundation
4600 Rickenbacker Causeway
Virginia Key, Miami, FL 33149

Research and organization focusing on the scientific study of the oceans and their role in life on Earth.

International Union for the Conservation of Nature and Natural
Resources (IUCN)
Avenue du Mont-Blanc
CH-1196 Gland
Switzerland

Promotes scientifically based action for the conservation of wild living resources.

Lake Victoria Research and Conservation Team
The New England Aquarium
Central Wharf
Boston, MA 02110

A network of research scientists, ichthyologists, and limnologists dedicated to saving Lake Victoria and its biodiversity.

Land Trust Alliance
900 17th Street NW, Suite 410
Washington, DC 20006

Provides legal and technical support to land preservation groups.

Massachusetts Audubon Society
S. Great Road
Lincoln, MA 01773

Regional society engaged in a broad range of ecosystem and special conservation activities.

National Audubon Society
950 Third Avenue
New York, NY 10022

Carries out research, education, and action programs to preserve wildlife and important natural areas and to protect the natural systems on which all life depends.

National Wildlife Federation
1400 16th Street NW
Washington, DC 20036-2266

Dedicated to creating an awareness of the need for proper management of soil, air, water, forests, minerals, plant life, and wildlife.

National Coalition for Marine Conservation
P.O. Box 23298
Savannah, GA 31403

Promotes public awareness of marine conservation issues and helps shape public policy.

Natural Resources Council of America
801 Pennsylvania Ave., SE, Suite 410
Washington, DC 20003

Provides information on actions taken by Congress and the Executive Branch as well as scientific data on conservation problems.

Natural Resources Defense Council, Inc.
40 West 20th Street
New York, NY 10011

Uses an interdisciplinary legal and scientific approach to monitor government agencies, initiating legal action and disseminating citizen information aimed at protecting endangered natural resources and improving the quality of the human environment.

Nature Conservancy
1815 North Lynn Street
Arlington, VA 22209

Committed to preserving biological diversity by protecting natural lands and the life they harbor; manages a nationwide system of over 1,600 nature sanctuaries.

New England Aquarium
Central Wharf
Boston, MA 02110

Public aquarium committed to the presentation, promotion, and protection of the world of water through its exhibits, and through education, conservation, and research programs.

New York Zoological Society
Bronx Zoo
185th Street & Southern Blvd.
Bronx, NY 10460

Promotes zoological research, increased public understanding of the environment, and wildlife conservation. Operates the Bronx Zoo, the New York Aquarium, the Osborn Laboratories of Marine Sciences, the Wildlife Survival Center, and Wildlife Conservation International.

Rachel Carson Council, Inc.
8490 Jones Mill Road
Chevy Chase, MD 20815

International clearinghouse for ecological information, focusing on issues of chemical, and especially pesticide, contamination.

Rainforest Action Movement
430 E. University
Ann Arbor, MI 48109

Focuses on protection and preservation of rain forests in Alaska, Oregon, Washington, Hawaii and tropical areas.

Rainforest Action Network
450 Hansome St., Suite 700
San Francisco, CA 94111

Works with and helps fund international organizations to preserve rain forests.

Rainforest Alliance
65 Bleecker St.
New York, NY 10012

Aims to link individuals interested in saving tropical rain forests.

Royal Botanic Garden, Kew
Richmond, Surrey TW9 3AE
England

A close collaborator with the IUCN in collecting and disseminating data on plants; the botanic garden contributes to public education as a living museum.

Seed Savers Exchange
3076 North Winn Road
Decorah, IA 52101

Dedicated to the exchange of seeds and the preservation of heirloom and old fashioned fruits, flowers, and vegetables. They have a seed bank as well as a living museum of the varieties.

Sierra Club
730 Polk Street
San Francisco, CA 94109

Focuses on legislation, litigation, public information, and recreation aimed at the exploration, enjoyment, and protection of the wild places of the Earth.

Smithsonian Institution
1000 Jefferson Drive, SW
Washington, DC 20560

Contributes to the public education through field investigations, the development of national collections in natural history and anthropology, scientific research, and publications.

Wilderness Society
900 17th Street NW
Washington, DC 20006-2596

Focuses on the preservation of the wilderness and wildlife and on the protection of America's prime forests, parks, rivers, and shorelands, and fosters an American land ethic.

Wildlife Conservation International (WCI)
New York Zoological Society
185th Street and Southern Blvd.
Bronx, NY 10460

International wildlife and wildlands conservation organization conducting field research and training programs around the world.

Worldwatch Institute
1766 Massachusetts Avenue NW
Washington, DC 20036

Identifies and analyzes emerging global problems and brings them to the attention of opinion leaders and the general public.

World Wildlife Fund
1250 24th Street NW, Suite 400
Washington, DC 20037

Principal private group in the United States financing conservation projects throughout the world. A primary collaborator on the World Conservation Strategy.

Xerces Society
10 SW Ash Street
Portland, OR 97204

Global conservation of insects and other invertebrates; instituted the Monarch Butterfly Project.

Contributors

David Ehrenfeld is Professor of Biology, Cook College, Department of Natural Resources, Rutgers University and Editor of the journal *Conservation Biology*.

Thomas J. Foose is Program Officer for the International Black Rhino Foundation and past Conservation Director of the American Association for Zoological Parks and Aquariums.

David Jablonski is Professor of Paleontology, University of Chicago, Department of Geophysical Sciences.

Les Kaufman is Chief Scientist and Head of the New England Aquarium's Edgerton Research Lab and organizer of a multinational group of scientists to save Lake Victoria.

Kenneth Mallory is Editor and Head of Publishing for the New England Aquarium and organizer of the biannual Lowell Lecture Series from which the chapters in this book evolved.

Norman Myers is Consultant in environment and development, Oxford, United Kingdom.

Ron Nowak is with the U.S. Department of the Interior Fish and Wildlife Service, Office of Scientific Authority, Washington, D.C.

Ghillean Prance is Director of the Royal Botanic Gardens, Kew, Richmond, Surrey, England.

Jim Williams is with the U.S. Department of the Interior Fish and Wildlife Service, National Fisheries Research Center, Gainesville, Florida.

Index